FOOD PRODUCT DESIGN

FOOD PRODUCT DESIGN

A COMPUTER-AIDED STATISTICAL APPROACH

RUGUO HU, Ph.D.

Nestle R&D Center, Inc.

TECHNOMIC
PUBLISHING CO., INC.
LANCASTER · BASEL

Food Product Design
aTECHNOMIC publication

Technomic Publishing Company, Inc.
851 New Holland Avenue, Box 3535
Lancaster, Pennsylvania 17604 U.S.A.

Printed in the United States of America
10 9 8 7 6 5 4 3 2 1

Main entry under title:
   Food Product Design: A Computer-Aided Statistical Approach

A Technomic Publishing Company book
Bibliography: p. 207
Includes index p. 223

Library of Congress Catalog Card No. 99-60758
ISBN No. 1-56676-743-1

献给我所爱的家

To my beloved family

Table of Contents

Preface

WITH the rapid development of computer technology over the past few years, more and more food scientists and researchers have begun to appreciate using computer-aided modern statistical methods to solve their food product and process design problems. As with any new concept, many questions have arisen that must be answered before its use can become widespread. Therefore, there is a need for a practical guidebook for food product developers so that they can take advantage of the progress of modern statistics and computer techniques.

This book is a tool for professional food engineers, technologists, scientists, and industrial personnel who want to expand and update their knowledge of computer-aided statistical experimental methods in the field of food product design, whether or not they are trained in statistics and computer applications. It has also been written for people engaged in applied research in universities, mainly those in departments such as food and agriculture, biological and chemical engineering, chemistry, and statistics. In addition, the book is designed as a reference guide for graduate students and upper-level undergraduate students who seek to explore the potential of statistical food product design with the effective aid of modern computer software. It also could be used as a textbook on computer-aided statistical experimental methods in a one-semester course in most applied curricula or, perhaps, could supplement the teaching material used in a two-semester course on basic statistical theory and statistical experimental methods.

Statistical experimental design is currently used as a quality control technique to achieve product excellence at the lowest overall cost. It is actually a powerful tool to optimize food products and/or processes, to accelerate food development cycles, to reduce research costs to facilitate the transition of products from research and development (R&D) to manufacturing, and to effectively troubleshoot manufacturing problems. The objective of this book is to familiarize readers with the method of statistical product design and encourage the application of this method in food product design, with the aid of widely available, modern computer software. In addition to the basic concepts of statistical food product design, this book presents the most effective statistical techniques and computer applications for trial design, modeling, and experimental data analysis. It includes numerous small BASIC programs with original codes and worked out practical examples from real research situations.

The first chapter introduces the common problems in food product design and the corresponding traditional and modern statistical methods used to solve them. The general steps in using the statistical approach to designing a food product are explained.

Chapter 2 discusses the concepts of the black-box modeling and the classification of independent variables into process and mixture variables. The essential difference between a factorial experimental design and a mixture design is explained using practical examples. A brief historical overview of the important contributors to statistical experimental methodology is also presented.

Chapter 3 presents the experimental techniques for product development in a food process. Task identification and screening of important variables, selecting the correct model form and experimental design, building a suitable model and testing its significance, analyzing the single and joint effects of variables, and product optimization and prediction are some of the main topics covered. The use of various kinds of common commercial software to solve problems of modeling, optimization, and prediction is shown in detail.

Chapter 4 deals with solving the problem of recipe or formulation modeling and optimization using statistical mixture experimental methods. The similar corresponding concepts used in a mixture system are discussed parallel to those presented in Chapter 3.

Chapter 5 tackles the most complicated, but the most common, problems in food product design: modeling and optimizing food systems, including recipe and process variables, based on statistical experimental methodology. A detailed practical example is provided to illustrate the technique and the many potential applications of statistical modeling and optimization.

A new aspect of computer applications, which uses the different kinds of significant models for an expert system, is presented in Chapter 6. The basic principles of an expert system are introduced in this chapter with a practical example based on those models presented in Chapters 3 and 4.

I am extremely grateful to many friends for their help in completing this book. In particular, I am indebted to Prof. Dr. W. Seibel, Dipl.-Ing. K. Seiler, and Prof. Dr. M. Kuhn, with whom I have had the pleasure of working on research problems with statistical approaches and who encouraged me to write this book. I particularly thank Dr. G. Ji and Prof. Dr. K. Specht who have reviewed the initial draft of this book. Their detailed and thoughtful comments on content and style were instrumental in the organization of this book. I thank Mr. Jay-Shake Li and Miss Min Ling Loh and the many colleagues at the German Federal

Center for Cereal, Potato and Lipid Research, and at the Nestlé R&D Center in New Milford, Connecticut, for everything they have done for me; also, thank you to the staff at Technomic Publishing Co., Inc., Lancaster, Pennsylvania. Finally, I express my special sincere appreciation to Mr. A. Osterwalder of the East Asia Institute, Bonn, Germany and Prof. Dr. Heinz Hesselfeld and Prof. She Shi Wang of the Jiangxi-OAI Joint Research Institute in Nanchang, China for their extensive support in completing this and other works.

I would like to encourage readers to comment on any errors and make suggestions for improvement. I hope this book will stimulate readers' interest in the possibilities and potential of computer applications and thus will encourage and enhance computer-aided statistical food product design.

DR. RUGUO HU

CHAPTER 1

Introduction

1.1 PROBLEMS OF FOOD PRODUCT DESIGN

IN the field of industrial food production and processing, the problem of optimization always arises, dealing with design, development, and quality management of food products. The attempt to develop a new food product or to improve the quality of a food product that largely meets the needs of consumers proves to be appealing and challenging. General questions about food product design could include:

(1) How can changes in one set of processing variables affect overall food quality? How are they correlated to each other? Which relationships are most important for food quality management?

(2) Which food raw materials and recipes should be used for production of an optimum quality food product? About 60–70% of the cost of most food products can be ascribed to the cost of raw materials.

(3) Which processing conditions must be used to achieve effective and successful production of a food product? Is an existing process running under control? Energy balances, likewise, are important, not because energy is a large cost factor in food processing, although it can be significant, but because correct application of energy may be the key to optimization of a process. Such important and unique processes as sterilization, pasteurization, freezing, drying, evaporation, baking, cooking, blanching, and frying all involve the addition or removal of heat to or from foods. Precise delivery of the correct amount of energy is important because both too much and too little energy transfer can be harmful.

(4) How can we determine the cause of food product defects and further control them? About 80% of all food quality defects are in part located in the design phase. One approach to track down potential problems and wrestle them under control before start-up of production is to design product quality right into the product.

(5) Most food products contain numerous ingredients, and manufacturing the product typically involves several different processing steps. How to choose the right combination of ingredients and processing steps which are important factors affecting different quality indices and desired quality characteristics of the food product?

(6) Are the recipes and the processing conditions currently used for an existing food product the best? How can we modify them to improve the food quality and/or to better meet the demands of the market? Once an optimum product has been developed, it will not remain so for all times. Usually, the quality of existing food products must be improved or modified with time to meet ever-varying consumer requests.

(7) What measures can be taken to increase production yield? How can we reduce the production cost and time of an existing food product?

(8) Is it possible to have a successful food product with cost-effective raw materials and procedures? How can individual production process be improved and wastes be minimized to improve the overall manufacturing process?

The demand for new food products and for high food quality requires extensive use of effective statistical experimental approaches by engineers and scientists. In the past, such an extensive use had been hindered due to a lack of proper training and, due to a lack of availability of appropriate computer hardware and software to implement experimental design and data analysis. This situation has now dramatically changed. Computer-aided statistical food product design will considerably improve the overall effectiveness of food research and development and enhance its application in the food research. In fact, any technique that could potentially improve the probability of a product's success in the marketplace is extremely attractive. If a particular product has optimal quality indices, such as sensory, nutritional, textural or physical, chemical and analytical attributes, it would lead to a better consumer acceptance. The attainment of optimized indices in a food product alone, of course, does not guarantee success in the marketplace, because factors, such as price, quality, brand, advertising, competition, loyalty, and economic factors, are additional major contributors to food product marketing success.

To achieve the objective of developing food products with optimal quality, we must first identify the properties and levels that are important to a specific food quality. It is important to keep in mind that optimal food products may not be able to be achieved uniquely through a combination of different product characteristics. To find out the optimum recipe and processing conditions for food products of high quality and high marketability is the critical issue. There are different methods to attain this objective, but hardly any of them can match the efficiency and the success of research and development activity in successfully supplying new and modified existing products. Among them, the statistical experimental methods such as Response Surface Methodology (RSM) are effective methods for food product design.

The theories of statistical experimental approaches have been developed for quite some time. However, they are not widely used for two main reasons. First, people in general are afraid of "statistics" and "mathematics" and, thus, avoid these methods and do not understand them. The other primary cause is the lack of suitable computer hardware and well-designed software, but this situation is rapidly changing. One of the objectives of this book is to familiarize the reader with statistical experimental methods and use them, with the aid of modern computer software, in food product design.

1.2 CLASSICAL ONE-FACTOR EXPERIMENTAL METHOD

Food process and recipe optimization is the essential task for food product designers. To accomplish this task, conventionally the One-Factor-at-a-Time Method, or One-Factor Method, is used, in which, at the first stage, only one variable is changed and tested at several levels more or less systematically, holding the other variables constant. In this way, the effect of this variable on product quality can be investigated, and a limited "optimum level" for this tested variable can be found by comparing the response results obtained. In the second stage, the optimum of a second variable can be determined. The level of the first variable is kept at its found optimum, whereas the effect of another variable is tested over several levels, while holding the rest of the variables at their constant levels. Thus, the optimal levels for each variable are found, and together they form the overall optimum for the specific product quality. This cycle is normally repeated until no significant improvement of the food quality can be detected.

This One-Factor Method is also known as the Trial & Error Method and has been used in the field of food product development for a long time. For example, salt, sugar, and smoke flavor are the major ingredients or variables responsible for most bacon flavors. Using this method to find the optimum flavor, one would change each of these variables (salt, sugar, and smoke flavor) individually, until a bacon formulation with the best flavor is achieved. In practice, the effect of salt would be tested by changing its level in the bacon while maintaining constant sugar and smoke levels and the product with the optimum flavor is selected. The optimal salt and smoke levels are then determined in a similar way. The "optimal levels" of salt, sugar, and bacon flavor would then be combined and considered as the overall optimum.

The obvious disadvantage of this experimental optimization method is its laborious experimental manner. The experimental points in the design space are not selected in a systematically structured way, and there is a low level of information extracted for experimental data collected. There is neither information about variable interaction effects nor an overview of the variables' behavior within the entire experimental space. The achieved optimum consists only of variable levels that were actually tested. This experimental method could lead

TABLE 1.1. Results of Five Trials of Proofing Time with $T = 25°$C.

Trial No.	(1)	(2)	(3)	(4)	(5)
Proofing time	60	90	120	150	180
Specific volume	1.78	2.10	2.38	2.25	1.93

Source: Cereal Foods World 3 (1987), 12: 857; reprinted with permission of the American Association of Cereal Chemists.

an inexperienced product developer to unreliable or even false optimal results, depending on if there are significant interaction effects between the variables, which is usually the case. In the above example, the change in sugar and/or bacon flavor levels would probably modify the optimal salt level; therefore, the "overall optimum" achieved might not be the true optimum.

Another obvious disadvantage of the Trial & Error Method is that it is inefficient and uneconomical. Usually, a large number of trials are required, especially if there are more than three variables to be investigated at multilevels, resulting in an expensive and time-consuming process. Additionally, a mathematical model would not be established by this method, so the overall relationship between the variables and the food quality (responses) could not be described quantitatively. Because food experiments in industry usually involve a large number of variables, use of designed experiments can cut the amount of costly experimentation substantially by reducing the number of trials needed to achieve a given objective.

To demonstrate the One-Factor-at-a-Time Method more clearly, an example is given below. The proofing time, t (min), and the proofing temperature, T (°C), are the two important independent variables influencing the specific volume, V (ml/g), of the loaf in the proofing process of breadmaking. The question here is: How can a proofing time and proofing temperature be found that will lead to production of a loaf with maximal specific volume? Based on pertinent literature and experience, a proofing temperature of 25–30°C was deemed

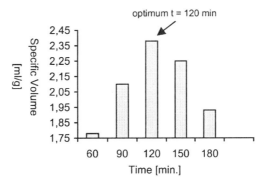

Figure 1.1 Optimum time by One-Factor Method.

TABLE 1.2. Results of Five Trials of Proofing Temperature with $t = 120$ min.

Trial No.	(6)	(7)	(8)	(9)	(10)
Proofing temperature	20	25	30	35	40
Specific volume	1.94	2.38	2.53	2.41	1.88

Source: Cereal Foods World 3 (1987), 12: 857; reprinted with permission of the American Association of Cereal Chemists.

reasonable. Following the approach of the One-Factor-at-a-Time Method, the proofing temperature, T, was held constant at 25°C, and the proofing time, t, varied from 60 to 180 minutes, with results that suggested that the best specific volume was obtained at 120 minutes (Table 1.1 and Figure 1.1).

To investigate the effect of proofing temperature, proofing time was held constant at 120 minutes, and proofing temperature varied from 20 to 40°C. Results suggested an optimal temperature of 30°C, which gave the maximum loaf specific volume of 2.53 ml/g (Table 1.2 and Figure 1.2). Thus, the optimum proofing time and temperature appeared to be 120 minutes and 30°C. In reality this conclusion is unfortunately inexact or incorrect. The true optimum levels should be about 32°C and 90 minutes as obtained through the statistical experimental method. The statistical results are plotted by using a contour graph in Figure 1.3, in which the optimization steps and results of the One-Factor Method are illustrated. In the graph the points numbered from (1) to (10) correspond to the "Trial No." in Tables 1.1 and 1.2 in the Trial & Error optimization method.

The reason that these different results were obtained is because significant quadratic and interaction effects exist between proofing time and proofing temperature on the loaf volume. The true relationship between these variables is shown in Figure 1.4. The term *interaction* should be explained here. If the effect of proofing temperature on the loaf volume (response) depends on the levels

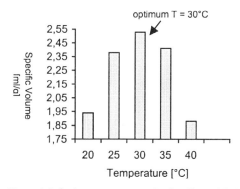

Figure 1.2 Optimum temperature by One-Factor Method.

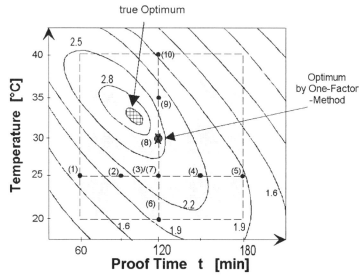

Figure 1.3 True overall relationship between the loaf specific volume and the proofing time (t) and temperature (T) (showed by contours); trial points of One-Factor Method [(1)–(10)].

of the proofing time, then there are interaction(s) between proofing time and temperature. In the One-Factor Method the interactive effect between any two independent variables is not considered, and the whole relationship is seen as described in Figure 1.5.

To offset the possible disadvantages of the conventional one-factor experimental methods, efforts have been made to develop new and more effective experimental methods based on statistical disciplines for purposes such as product

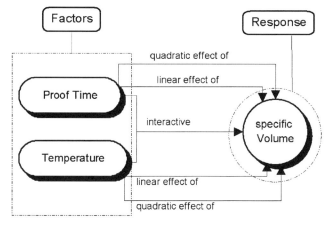

Figure 1.4 True relationship in the system.

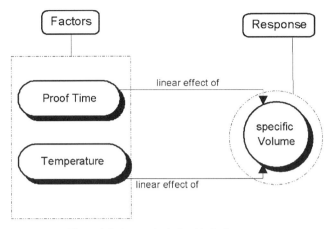

Figure 1.5 Assumed relationship in the system.

design, process modeling, and optimization. In these modern methods the main and the interactive effects of every independent variable are considered and checked, that is, overall relationships in the whole experimental space are determined and described in the form of mathematical models. The total number of experimental trials is reduced considerably compared with the classical Trial & Error Method, and there is no significant limiting of information value or precision with the results obtained. With the current availability and use of powerful personal computers and user-friendly statistics-oriented software, complicated calculations and graphics that proved almost impossible in the past can now be done with great ease.

1.3 GENERAL REMARKS ABOUT STATISTICAL METHODS

An alternative and modern approach of solving food product design problems is through the use of a computer-aided statistical experiment method. By using supporting data from a structured experimental design, the whole problem is modeled systematically. It is then possible to analyze and determine the relationship between variables and to optimize the recipe and/or the production processes as directed by the significant model. Statistical modeling techniques were originally developed by mathematicians and statisticians, yet its successful application in solving food development problems was limited because of the lack of suitable statistical software and widespread use of computers. Recently, there has been a rapid growth in the use of powerful personal computers and of well-designed computer programs available to perform often laborious modeling and analysis calculations. This has opened a new opportunity for application of statistical experimental methodology in food product design.

Statistical experimental methodology can be considered a quality control technology that achieves food product development and product excellence at the lowest possible overall cost. With the application of computer-aided statistical product research and development (R&D), the development cycle can be shortened and accelerated, research costs can be reduced, the transition of products from R&D to industrial production is easily facilitated, and manufacturing problems can be solved through effective trouble shooting procedures.

1.3.1 IMPORTANT CONCEPTS AND TERMS

Before discussing statistical food product design any further, some important basic concepts should be explained. These terms are used throughout the book and will not be explained again.

- Independent variables or variables, more commonly known as factors, are parameters or characteristics, normally including ingredients and processing conditions during the food-manufacturing process, which have an effect on product quality and can be varied within a definite range of interest. Temperature, time, pressure, pH value, moisture content, amount or type of ingredients, enzyme activity, agitation intensity, machine type, and controllable machine parameters are some of the more common factors in food production.
- Dependent variables, responses, or target parameters are the important measurable food product quality indices of interest and are influenced directly or indirectly by different factors. These are important characteristics in deciding quality of the food products. The issue of what responses are considered important can be decided by one individual or a project team. Sensory score, nutritional value, physical or rheological properties, chemical composition, microbiological and hygienic characteristics, viscosity, color, yield, cost, and shelf-life are some concrete examples of responses.
- Test levels or levels are the values or quantity of the factors selected to be tested in the experimental design. The difference between the maximal and the minimal test levels for a variable is known as the wide span or level span of this variable.
- An experimental design maps out a combination of factor levels for an experimental plan.
- When the factor level settings for two factors in an experiment are uncorrelated, that is, when they are varied independently of each other, then they are said to be orthogonal to each other.
- A model is a mathematical equation or function that describes the relationship between the response values and different factors quantitatively. It can be used to predict response values for different factor levels, or vice versa (optimization). Empirical polynomial equations are usually used for modeling.

- Interaction (interactive effect) between two variables means that the effect of one variable on the target parameter depends on the level of the other variable. Similarly, there may be interactions between more than three variables, but these are usually insignificant and physically unexplainable, thus normally omitted in a model.
- A trial, a run or an observation is a single test in a designed experimental design. The whole experimental design is constructed of a series of trials. A single trial is one support point for model building.

1.3.2 STATISTICAL EXPERIMENTAL METHOD

The statistical experimental method can be summarized as modeling a food process and formula based on data collected from designed experimental designs. The model built is then used to investigate the single and the overall relationship between all variables and food quality indices through numerical or graphical methods, and finally to make predictions and optimizations of the food quality. With this systematic approach, food scientists and engineers are able to extract the greatest amount of meaningful information from the fewest number of experiments and at the lowest cost.

To design an experimental plan includes selecting the experimental points (support points) in a definite manner based on statistical principles and real situations, with which the process or the recipe can be represented in the form of mathematical models with high precision. Designing an experimental plan can be somewhat likened to the wheels on a car:

- First, to enable a car to run it needs at least three wheels (support points), but four wheels are better because of higher driving stability. More than four wheels could be thought unnecessary.
- Second, the four or three wheels must be installed at the appropriate places under the car to achieve high stability for the car. Obviously, a car with four wheels on one side cannot run. Similar considerations apply in the selection of the support points in a statistical experimental plan. Only certain trials should be selected for testing.

Figure 1.6 demonstrates the importance of proper selection of support points (trials) in a one-factor problem. The curve shows the true relationship between a variable X and a response Y. The experimental plan with the ☆ support points has a much better distribution than that of the one with the • support points, even though there are more of the • support points in the experimental plan. A well-designed experimental plan with fewer support points can supply more and better information than poorly designed plan with more observations.

With the suitable selection of a minimal number of experiments or observations (sufficient number of wheels with correct installation on the car), the food system can be described effectively through mathematical procedures. This step is also called modeling, in which an equation is postulated or arranged to

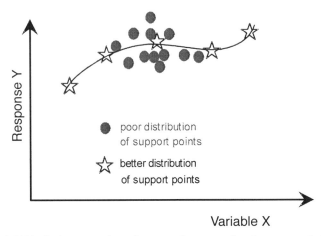

Figure 1.6 Distribution comparison of two sets of support points for one true relationship.

represent the response. However, the designed experimental plan must correspond to the model to be fitted in terms of regression. The significance of the model must first be tested before it is used for food product design purposes such as product optimization. The statistical model testing ensures that the fitted equation is reliably accurate and based on which predictions and optimizations can be carried out effectively and precisely. Using a significant model, the food product developers are able to examine whether the changes made in any of the variables have a critical effect or not on the response.

1.3.3 ADVANTAGES OF STATISTICAL APPROACH

In contrast to traditional techniques of food product development in which only one variable is tested at a time while all other variables are held constant, statistically designed experimentation, or statistical experimental design, offers a simple yet comprehensive approach to developing quality food products and efficient processes. Experimental design offers several advantages over conventional investigation methods, which are summarized as follows:

(1) The significant variables among all the possible independent variables and interactions between variables can be determined by using a statistical experimental design.

(2) Several independent variables and their responses are studied systematically at the same time. With a statistical experimental design the relationship between all variables and the response in the whole research space is determined and continuously described through a mathematical model.

(3) Effective information extraction from fewer trials and an optimal ratio of time and resources consumption to the information obtained is achieved

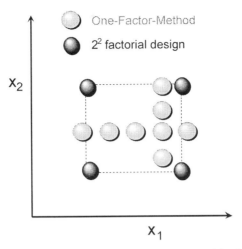

Figure 1.7 Comparison of trial distributions of a factorial design and the One-Factor Method.

with a statistical experimental design. Figure 1.7 demonstrates both the distributions of the support points of the One-Factor Method (with eight trials) and of a typical 2^2 factorial design (with four trials). Obviously, the trials of the factorial design can support information extraction from a broader range.

(4) Prediction of responses at all possible combinations of independent variables can be made and vice versa (optimization) with a statistical experimental design.

(5) In contrast to the One-Factor Method, "true" optimums can be identified reliably through numerical or graphical analysis procedures.

1.3.4 STEPS TO STATISTICAL FOOD PRODUCT DESIGN

The statistical approach of food product development entails eight important sequential steps. In Figure 1.8, these steps are illustrated in the form of a flowchart for clarity. Each step is described in detail as follows.

1.3.4.1 Step 1: System Identification

To achieve the objective of developing food products with optimal quality or of improving the quality of existing products, first, we must identify the critical or specific food properties that are important to this product. This implies that the purpose of the study be clarified, that the responses and their corresponding variables (factors) be selected and identified, or, in a word, that the system under study be defined. Most food products contain numerous ingredients, and their manufacture typically involves several different processing steps such

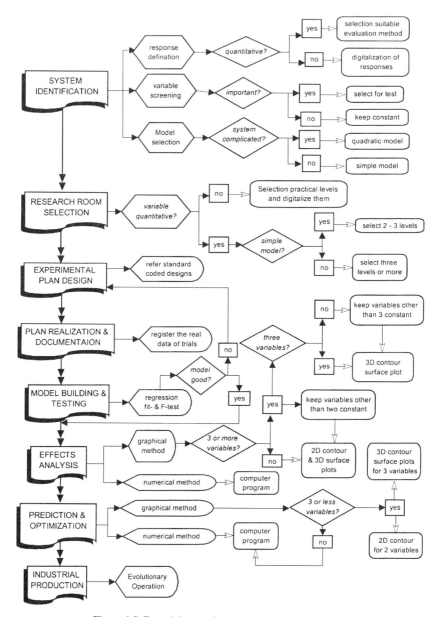

Figure 1.8 Essential steps of statistical food product design.

as blending, heating, and cooling. All ingredients and their processing steps are surely not equally important to different product quality indices. Thus, we must first try screening and listing all the relevant variables to compile qualitative information about which specific variables affect the targeted product qualities (responses). Next, we must analyze and determine the relationship between the responses and all important independent variables. It is helpful to present the results in the form of a flowchart for the sake of convenience and understanding.

Independent variables can include ingredients or processing conditions and may be qualitative (e.g., types of flours or enzymes or machine types) or quantitative (e.g., 5% vs. 10% sugar content). Dependent variables or responses are the properties of the end product to be measured, sometimes including sensory scores, nutritional value, texture, viscosity, microbiological stability, and shelf life.

Special effort and care must be taken during this step. If important variables are missing or remain undetected during this stage, they will not be shown in the statistical model, because regression computations will not invent them automatically. The optimized recipe and processing conditions will be accordingly a failure. It should be kept in mind that the decision about which responses and factors are important could be different, depending on whether the opinion of an individual or of a food product development team is used. This illustrates how important for the food product researcher to have an intimate understanding of materials, processes, and the food-marketing objectives. To be able to fully recognize the problems just discussed is the initial requirement for a successful product design project.

1.3.4.2 Step 2: Research Range Selection

The independent variables to be controlled are selected from experience or based on research results obtained from literature or preliminary experiments. These variables must be important and relevant to the production process and the food product quality because they can have either a direct or an indirect influence on food quality. Suitable test levels for each (independent variable) are chosen to be meaningful and realizable in practice. The test levels of all the factors then build a "room," or "space," in which the relationship between the factors and the responses will be determined quantitatively and described with a mathematical equation.

It is sometimes not so easy to select reasonable experimental levels for each variable. The level span between the upper and lower values of each variable must not be too large; otherwise, the model to be built might not be adequately significant. On the other hand, the level span of one variable should also not be too small; otherwise, this variable itself could lose its significance in the model, and the total reliable information obtained from the model would be reduced.

1.3.4.3 Step 3: Experimental Plan Design

Based on the number of variables and the number of test levels for each variable, a statistical experimental plan is constructed. There are a number of standard coded statistical experimental plans from which a choice can be made (see Appendix). Several modern computer programs supply designs as well. In step 2, reasonably wide spans of variable levels were determined. The number of test levels of one variable depends on the statistical experimental plan chosen and the real relationship of this variable to the response. In fact, what kind of an experimental plan should be used depends on the model form, which is, in turn, decided by the complexity of the system of the food-manufacturing procedure. In principle, experimental plans with more than two test levels should be adopted to an obvious nonlinear process (usually for optimization) to allow determination and modeling of interactions and/or curvature effects of variables.

All experimental programs have time and cost constraints, but given these constraints, not all experiment designs are equally powerful. Given the same number of variables and test levels, there may be several suitable experimental designs that can be used. Many statistical techniques can be used to generate different designs to meet specific demands. It is normally sensible to select experimental designs containing a smaller number of trials. During the selection of designs, it is important to make sure that there are enough experimental runs so that the experimental error and the goodness of the model fit can be tested. Normally, the degree of freedom for the experimental observations must be obviously higher than that of the model, which equals the number of terms in the model including the constant term.

1.3.4.4 Step 4: Plan Realization and Documentation

The trials in an experimental design should essentially be conducted in a randomized order so that their correlation with time may be minimized. Such randomization is not always feasible or practical. Variables, such as temperature or machine type, are often quite difficult to vary. In such cases, the trials in a statistical experimental plan can be arranged so that difficult variable changes are minimal.

All trials in a randomly ordered experimental plan must be performed. Every test level of a single variable should be held at the planned value as closely as possible. However, in reality, there are always deviations between the set values and the true values. Therefore, the true experimental data from each trial, including both the variables and responses, must be documented in detail for final modeling. Recording the data meticulously and accurately in full agreement with the practical experimental results is the essential prerequisite for successful food product development. Remember that false data can be modeled as well, but an effective and successful product optimization cannot be expected on the basis of such a "model."

1.3.4.5 Step 5: Model Building and Testing

In this step, a model for each response (if there is more than one target response) is built and evaluated in accordance with all actual experimental data in terms of regression and corresponding statistical tests. A model is actually a mathematical equation that describes the relationship between the food quality indices (response) and its independent variable. It can be used to predict response values for different variable levels or vice versa. To "build a model" means nothing more than to find the most suitable coefficients of each term in the equation so that it fits and describes the observations most precisely. It must be mentioned that a different model should be set up for each response.

Developing a model is perhaps one of the most challenging aspects of statistical food product design due to complicated calculations. Fortunately, nowadays it can be done easily by using suitable computer software. The single most important thing for the product developer to do is to examine the modeling statistics and make sure an appropriate model has been finally created. Stepwise multiple regression procedures, including two alternatives of forward or backward regression, are usually used to fit the experimental data. According to the forward method, the independent variable that maximizes the model fit or contributes most to the prediction of a dependent variable (response) is selected first. New variables are added to the regression equation in order of their contribution to the prediction of the dependent variables, provided that they contribute at a specified level of significance. The backward procedure starts with a full model and eliminates variables one by one from the model. At each step, the variable with the smallest contribution to the model is deleted. The action in these two regression methods is normally performed according to a criterion that can be specified by the user. A series of commercial computer programs providing these stepwise multiple regression procedures is available.

The significance of the model must be checked according to the F-test and the lack of fit, or R^2, before it can be used further. These tests are statistical criteria to examine the correctness or the effectiveness of the model and to estimate the experimental error. For example, a large R^2 value indicates a workable regression equation. Thanks to the availability of computer programs, these tests can be performed quickly and with great ease. Almost all modern computer statistical software supplies a variety of diagnostic test results during the process of model building and often gives additional detailed information about the regression calculation. This information sometimes include the analysis of variance (ANOVA), the F-test, the residual analysis, the standard error, the percent variation accounted for by the model, and the coefficients of variation.

Once the model is proved adequately significant, it can be used further for analyzing and predicting the effect of any combination of evaluated factor values. The factor values can be the tested levels or any values between the minimal and maximal levels. If the desired significance level of the model is not assured, then either the form of the model should be changed, usually through transformation of variables, and step 5 should be repeated. Sometimes additional trials should be included in the original design for building of a model of higher degree, therefore taking us back to step 3.

As mentioned in the first step, if important variables are not identified and controlled in the design and not present in the statistical model (model-related statistics may show evidence of missing variables), the regression computations will not invent them. Correct results cannot be achieved, even if the model built seems "significant!" Therefore, it is extremely important that all the relevant variables have been accounted for in step 1.

1.3.4.6 Step 6: Analysis of Variable Effects

One objective of using a statistical experimental approach in food product design is to exactly identify those variables or combinations of variables that are important to food product quality. If the model proved to be statistically valid, it can be used to analyze and make predictions on the effect of different levels of variables on food product characteristics within the area studied. A significant model can contribute greatly to the development of an organized set of information about the relationship between food quality and variables and thus make an important contribution to the efficiency of the product development process.

Based on the model, numerical or graphical [2-, 3-dimensional (2D, 3D)] methods are used to determine the effects of variables and their possible interactions on target food quality indices. In a graphical method, the model is simplified to two or three variables with the other variables fixed at constant levels. Then 2D X-Y- charts and contours and 3D contour surfaces are drawn according to this simplified model. These plots visualize the relationships between the variables and the characteristics of the food product and are effective simple approaches to investigate the effects of variables. Especially useful for the layman, graphics supply direct visual information and helps to discern the relationships between factors and responses easily. By studying these graphics, regions of interest for food product design can be identified, which suggest further investigations or optimizations.

1.3.4.7 Step 7: Prediction and Optimization

Recently, the term *optimization* has become popular in many sectors of the food industry. It appears to have different meanings, depending on what one is concerned with, including product development, manufacturing, quality management, or even advertising and marketing. Optimization in food product design can simply be described as a procedure for developing the best possible product in its class. In other words, optimization means establishing the recipe and the processing conditions for production of the best of all possible food products. The intention of optimization is to provide a more precise map of the path that has the highest probability of leading to a successful food product. Although optimization is still part science and part art, it offers a simple yet comprehensive approach to developing quality food products efficiently.

To find the optimum recipe and/or processing conditions for the best food products, numerical or graphical methods based on the model can be used. Optimization is actually a pure mathematical calculation, in which the desired characteristic of the food product (response) is defined and targeted so that the optimal recipe and process conditions can be calculated or vice versa. The desired food product attributes are often a maximal or minimal value. In case there is more than one response to be considered, optimization is usually compromised, to the extent that an overall consideration of all the food quality indices is required. The most important response may be targeted more than minor ones. Given weighting factors and using desirability are an effective way to deal with such optimizations.

The principle of the graphical optimization approach is similar to the graphical method described in step 6 and is discussed in detail in Chapters 3 and 4. The optimization of all aspects of a product is the ultimate goal in food product development. It is important to know that more than one single optimal product may exist and that an optimum response may probably be achieved with different combinations of important independent variables.

1.3.4.8 Step 8: Industrial Production

To develop a mathematical model is not the final goal of food product design. There has been an understandable tendency in many areas of industrial food research to concentrate on the manufacture of edible products instead of collecting useful data. This tendency arises from confusions between the task of process research and that of product development.

It is obvious that the last thing one should do in statistical food product design is to start manufacturing the calculated optimum food directly. Before the optimization results are transferred to industrial production, their correctness must be tested and checked. The advantages of the new products developed should be outlined and the possible difficulties and potential problems during production according to the model should be taken into account and noted. For a successful transfer of optimum results from a laboratory scale to an industrial scale without disturbing the production, the technique of Evolutionary Operation (EVOP) is usually used to relocate the optimum condition for an industrial production apparatus.

1.4 SOFTWARE: A STEP TOWARD SIMPLICITY

Despite its potentially significant economic benefits, statistical food product design or statistical experimental approaches are often dismissed as "too complicated to do," or as "too difficult to understand." One barrier to using the statistical food product design has been lack of user-friendly, interactive software or lack of adequate knowledge about actual computer software. The development of computer technology has made a tremendous impact on all aspects of modern society. In some respects, the advances in computer applications have aided in the investigation and optimization of complex systems. However, to take advantage of the capabilities of computers, the underlying requirement is to mathematically represent (model) the system under consideration. The data and graphics processing power of computers can play a vital role in the food product design scheme. Computer graphics have been successfully applied to science (e.g., geography, geology, and space exploration) and engineering (e.g., design of cars, boats, and space ships), and can be valuable to food technologists as well.

In this book the computer will be used as an important aid to help nonstatisticians successfully implement designed experiments, build a mathematical model, analyze the model, and interpret it as well as make optimizations. Using popular software to simplify or even automate the conventional statistical approaches will enhance the independence and the flexibility of the food product developer during his work, taking only about one tenth of the time required by manual procedures. Following, listed alphabetically, are some popular software programs that may be useful in statistical food product design. The mail, E-mail, and Internet addresses of these software manufacturers or sale representatives are also given.

APO
SYSTEGRA GmbH
Frankfurterstr. 33-35
D-65760 Eschborn
Germany
Tel.: (+49) 6196 70 34 37
Fax: (+49) 6196 70 34 10

AXUM
TriMetrix®, Inc.
444 NE Ravenna Blv.
Suite 210, Seattle
WA 98115
USA
Tel.: (+1) 206 527 1810
Fax: (+1) 206 522 9159
URL: http://www.trimetrix.com

COREL CHART
© Corel Cooperation
1600 Cailing Avenue, Ottawa
Canada
Tel.: (+1) 613 728 8200
Fax: (+1) 613 761 8049
URL: http://www.corel.com

MS-EXCEL
Microsoft Cooperation
Tel.: (+1) 800 426 9400
URL: http://www.microsoft.com

HARVARD GRAPHICS©
Software Publishing Corporation
111 North Market Street
San Jose, CA 95113
USA
Tel.: (+1) 408 537 3000
Fax: (+1) 408 537 3500
URL: http://www.spco.com

©MICROGRAFX CHARISMA
©Micrograx, Inc.
1303 E. Arapaho Road
Richardson, TX 75081
USA
Tel.: (+1) 800-676-3110
Fax: (+1) 972-234-2410
URL: http://www.micrografx.com

MODDE
Umetri AB
Box 7960
S – 907 19 Umeå
Sweden
Tel.: (+46) 90154840
Fax: (+46) 90197685
URL: http://www.umetri.se

SAS®, JMP
SAS Institute Inc.
ASA Circle, Box 8000
Cary, Nc 27512-8000
USA
Tel.: (+1) 919 467 8000
Fax: (+1) 919 469 3737
URL: http://www.SAS.com

SIGMASTAT, TABLECURVE*TM*
Jandel Scientific
2591 Kerner Blvd.
San Rafael, CA 94901
USA
Tel.: 415 453 6700
Fax: 415 453 7769
E-mail: Sales@jandel.com

LOTUS 1-2-3 & LOTUS IMPROVE
Lotus Development Corporation
55 Cambridge Parkway
Cambridge, MA 02142
USA
URL: http://www2.lotus.com

MINITAB
Minitab Inc.
3081 Enterprise Drive
State College, PA 16801-3008
USA
Tel.: +1-814-231-2682
Fax: +1-814-238-4383
E-mail: techsupport@minitab.com

ORGIN
Microcal Software Inc.
One Ronudhouse Plaza
Northampton, MA 01060
USA
Fax: (+1) 413 385 0126
E-mail: compuserve 100144,347
URL: http://www.origin.com

SCIENTIST™ for Windows
MicroMath Scientific Software, Inc.
2469 E. Fort Union Boulevard
Salt Lake City, UT 84121-0550
USA
Tel.: (+1) 800 942-6284
Fax: (+1) 801 943-0299
URL: http://www.micromath.com

S-Plus
MathSoft, Inc.
101 Main Street
Cambridge, MA 02142-1521
USA
Tel.: 617-577-1017
Fax: 617-577-8829
URL: http://www.mathsoft.com

SPSS
SPSS Inc.
444 N. Michigan Avenue
Chicago, IL 60611
USA
Phone: 1 (800) 543-2185
Fax: 1 (800) 841-0064
URL: http://www.spss.com

STANDFORD GRAPHICS
Visual Numerics^{TM}
Zettachring 10
D-70567 Stuttgart
Germany
Tel.: (+49) 711 13287 0
Fax: (+49) 711 13287 99
URL: http://www.vni.com

STATGRAPHICS
Manugistics, Inc.
2115 East Jefferson Street
Rockville, MD 20852-4999
USA
Tel.: 301-984-5000
Fax: 301-984-5370
URL: http://www.statgraphics.com

STATISTICA
StatSoft® Inc.
2325 E., 13th Street, Tulsa
OK 74104
USA
Tel.: (+44) 918 583 4149
Fax: (+44) 918 583 4376
URL: http://www.StatSoft.com

SYSTAT
SYSTAT, Inc.
1800 Sherman Avenue
Evanston, IL 60201
USA
Tel.: (+1) 708 864 5670
Fax: (+1) 708 492 3567
URL: http://www.spss.com/software/science/systat

1.5 A BRIEF HISTORICAL PERSPECTIVE

Food product design through statistical experimental approach is not new. From about the 1920s the statistical experimental method has been explored and updated by numerous statisticians. For at least 50 years following its conception, the methodology was largely underused by the food industry. Only since the late 1980s have statistical experimental approaches resurfaced for use in both food quality control and food product development, seemingly as a "new" technique in experimentation.

In chronological sequence the names of some important statisticians and their relevant contributions are given below. It is impossible to include all the important authors during this period for lack of space. For more information see the Bibliography.

(1) R. A. Fisher: In the 1920s, the Englishman Sir R. A. Fisher developed the foundations for statistical design of experiments and data analysis methods

such as analysis of variance. He applied this method successfully in agricultural research.

(2) G. E. P. Box, W. G. Hunter, J. S. Hunter, O. L. Davies, and J. Kiefer: These four statisticians are (among) the most important contributors to the theoretical development and practical application of statistical factorial experimental methodology. They have published numerous important papers in the 1960s and 1970s. The statistical experimental methods have developed with such diversity and flexibility that they can be applied to solve all kinds of problems in the field of science and technology. Optimization techniques were also developed and applied in chemical and allied industrial research.

(3) H. Scheffé and J. A. Cornell: Scheffé published his pioneering article "Experiment with mixture" in 1958, in which simplex-lattice designs and the corresponding polynomial models were introduced. In 1963 he introduced an alternative design to the {q, m} simplex-lattice design. Cornell reviewed almost all the published statistical articles on mixture designs and models in 1973 and published his book in 1981 entitled *Experiments with Mixtures* in which the basic theory of mixture experimental methodology was presented with numerous practical application examples.

(4) G. Taguchi: Genichi Taguchi modified in 1975 the existing statistical methods and developed a unique robust experimental design for off-line quality control. The three most important concepts of quality loss functions, signal-to-noise ratio and orthogonal arrays are used in this design method, which are set apart from traditional quality control procedures and industrial experimentation.

Making use of computer technology, statistical experimental methods during the 1980s became increasingly widespread, leading to a rational process for solving problems in all fields of research and product development.

CHAPTER 2

Problems of Food Product Design

2.1 SYSTEM IDENTIFICATION AND DEFINITION

THE first step in food product design is to identify the task and define the system for study as described in Chapter 1. The system must be analyzed in detail, and all the relevant and targeted product quality characteristics must be well defined. Selecting the quality indices of the food product to be developed is a challenging task. Basically, these quality indices must be factors that directly or indirectly influence consumer acceptance, food nutritional value, and the product cost. The technical features of an optimal food product designed must fall within the practical constraints of doing business. In addition, the objective quality indices of the food products should be observable and measurable. Sensory evaluations, nutritional values, physical tests, and chemical analyses using standard laboratory procedures can be chosen to identify those indices that are most likely to affect the overall quality of the food products. These standard analytical procedures are usually specific and appropriate for the targeted food product, and the data derived can be expressed or transformed numerically for modeling.

If a set of product quality indices selected do not include those characteristics important to food quality, then the results obtained from the optimization process will not be exact and sometimes even inappropriate. The team involved in identifying and selecting the targeted product quality indices should include specialists who understand this food category and are familiar with the properties and the ranges in which they can occur. Subject selection and training techniques should be used to develop the most reliable and valid set of quantitative, descriptive results.

Specific features associated with the product must be included during the identification of the product quality indices. Shelf life might be particularly required if the product is shipped and stored or consumed at high environmental temperature and humidity in tropical regions. For sensory quality, the environmental conditions, such as time, temperature and humidity at which the product is consumed, the many different ways that a consumer uses this food, the consumers (children or adults or elders) of the product, the region where the food is to be sold, and so on, should be considered. Children may opt for a sweeter and more aromatic version of the same product than adults do. All this should be considered to obtain meaningful data of sensory evaluations.

The system identification and definition include the identification and selection of independent variables (factors) besides the quality indices (responses) definition. The important and significant variables must be screened and selected for test. Making a list of all the possible factors—usually in a flowchart—will help the food product developer to understand the system more clearly. During this stage, some basic considerations listed below may be helpful:

- Which independent variables should be tested and their effects determined?
- Which variables should not be tested and be kept constant?
- Which levels of variables should be selected and tested in study?
- How should a qualitative variable be expressed numerically?

For a successful screening of independent variables and the selection of their test ranges, which are defined by the combination of all the variable levels, the food product researchers must refer to the literature, his experience, and some preliminary experiments. About 10–25% of total working time at this stage of system identification and experimental plan design is eventually adequate.

If important variables are not detected, then they will not be considered in the regression model, and correct optimization results will not be achieved; occasionally, even the resulting model may be apparently "significant." During variables identification we must check if they are duplicated. For instance, if the amount of water added to the dough has been chosen as one of the variables during breadmaking, then the moisture content of the dough cannot be taken as another independent variable of the system. One must be particularly careful with assumed variables. They should be checked and may be discriminated through some simple designed experiments.

Many variables have direct or indirect effects on the food quality. However, their effects cannot be evaluated simultaneously in a single statistical experimental plan. To enable an effective and practicable food product development, the test ranges have to be limited and minimized by reducing the factor number to an acceptable level. All variables can be usually classified into two categories according to their importance with respect to the desired food quality indices. The important variables should be controlled during product design, whereas those of minor importance may be held constant and not tested. Such an operation is the basics of the Black-Box Modeling.

2.2 BLACK-BOX MODELING AND INPUT-OUTPUT CONSIDERATION

Food products usually contain numerous ingredients (e.g., cereal flour, sugar, salt, yeast, enzymes, flavoring agents, and pigments), and their manufacture typically involves several different processing steps (e.g., blending, heating, cooling, stirring, and packaging) that are not equally important to the desired quality characteristics of the food product. Some combinations are beneficial for preservation, but the others are good for cost or nutritive value. To simplify the procedure of food product design, the Black-Box Model provides an easy analysis, identification, and modeling of a food system. The independent variables (factors) can be considered as input variables and the food quality indices (responses) as output variables. The other subprocesses and minor variables are held unchanged during the study and assumed as nonvariable—just like being stuck in a magic black box. The direct controllable independent variables in input are chosen for test, whereas the indirect variables or subvariables held constant. An output variable is considered as shown in the following equation:

$$\text{Output Variable} = F \text{ (Input variables)} \qquad (2.1)$$

This concept simplifies the system as well as the model built on it. In this Black-Box Model, the effect of the variables on the responses can be easily estimated and summarized in terms of statistical experimental methods. The relationships among the independent variables and the responses can be described in terms of mathematical functions. The following example illustrates the black-box concept and its strategy in identifying a food product design problem.

Food extrusion is a popular and modern food-processing technique, and many factors will influence the extruded food quality. Variables that may have direct or indirect influence on the extrudate characteristics are:

- sorts and homogeneity of raw materials
- raw material formulation, especially the mass moisture content
- extruder type
- length:diameter ratio of the extruder barrel
- extruder barrel temperature profile
- die temperature and pressure
- screw configuration
- screw rotation speed
- specific mechanical energy (SME) dispersion by the screw(s)
- die shape and dimension
- pressure profile in the barrel
- mass pressure in front of the die

- feed rate
- shear rate profile
- residence time and its distribution (RTD)
- chemical changes in the barrel
- room temperature and relative humidity
- stability of the electricity voltage
- skill of the operator(s)

These factors all influence directly or indirectly the extruded food qualities—sensory, functional, or nutritional properties. In practice, it is impossible to test the effect of so many variables on the extrudate quality at the same time. They must be classified and selected for further tests only the most interesting and controllable variables. The other (minor) variables, such as extruder type, room temperature, and room humidity, may be considered stable and not be taken into account in the extruded product development. Some of the other variables should also be held constant to reduce the number of variables to be tested. This will further reduce the complexity of the system. Figure 2.1 shows the results of a system analysis during extruded food product design. Some independent variables are controllable working parameters:

- sorts and formulation of raw material
- barrel temperature or its profile
- screw configuration
- screw rotation speed
- die shape and dimension
- extruder type, especially barrel length
- feed rate
- room temperature and humidity

The other independent variables, such as shear rate, SME, RTD, pressure profile, and chemical changes in the barrel, are not directly controllable. In fact, they are functions of these controllable variables and have further effects on the nutritional and sensory quality indices of the extrudates. As shown in Figure 2.1, during extrusion cooking, only the most important and directly controllable working parameters, such as barrel temperature, raw material formulation, screw rotation speed, and feed rate, would be selected and tested. All the other variables are held unchanged or disregarded as if they were stuck in a magic black box. Actually, more complicated relationships exist between these variables. For example, the shear rate will be affected by the sort of materials and formulations, the screw rotation speed, the temperature and the pressure profiles, which further influences the SME; the SME will directly be affected in a subprocess by the screw rotation speed as well as the flow properties of the material. The formulation of the raw material will have a significant effect on the nutritional quality of the extrudate, which moreover would be modified in a subprocess by the extrusion variables such as barrel temperature and shear

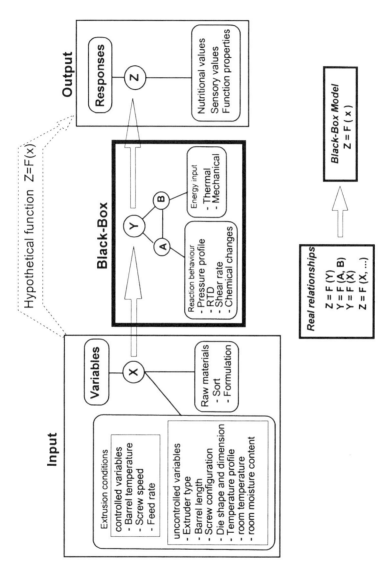

Figure 2.1 Analysis of extrusion problems and the black-box concept.

27

rate. In the Black-Box Model, these interrelationships are not considered. The effective model(s) to be built for this system is given in Equation (2.2):

$$\text{Responses} = f\ (\text{Raw material formulation, Barrel temperature,}$$
$$\text{Screw speed, Feed rate}) \tag{2.2}$$

The responses can be nutritional values, sensory evaluation data, or texture or function properties of the extruded food products. After a clear identification and definition of the system, or in other words, after the independent variables and the desired product quality indices have been chosen, we can go on to the second step of food product design, namely, to establish the statistical experimental plan and find the quantitative relationships among variables and objectives. There are, in fact, different ways of doing this depending on whether research focuses on food formulation or extruder operating parameters. This is described in Section 2.3.

2.3 PRODUCTION PROCESS PROBLEM

In food product design there are two different kinds of system problems—process and mixture problems—that should be dealt with by different statistical experimental methodologies. In the process problem, all the independent variables are not related to each other but are orthogonal to each other. The change of one variable is not restricted by another variable. Geometrically, the lines representing these variables meet at right angles (Figure 2.2). In breadmaking, for example, the temperature of the baking oven can, in principle, be chosen without any influence on the setting of baking time. Of course, only suitable settings of oven temperature and baking time can lead to desired bread-baking results. To solve a process problem, these statistical experimental designs used should contain no or few correlations between the independent variables, so that their natural or original properties of "independence" can be retained. These

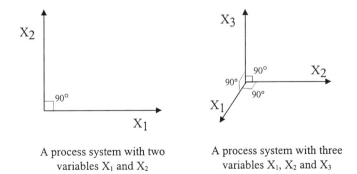

A process system with two
variables X_1 and X_2

A process system with three
variables X_1, X_2 and X_3

Figure 2.2 Illustration of the orthogonal properties of independent variables in a process system.

kinds of statistical experimental designs are usually factorial experimental designs or designs derived from it. More details of this approach are discussed in detail in Chapter 3.

2.4 RECIPE PROBLEM

The recipe is one of the most important factors leading to successful food products. A recipe usually includes several ingredients, which have different effects on specific food quality. To study these effects is the prerequisite for being able to choose the optimal recipes.

Many food products are manufactured by mixing two or more ingredients. In bread and cake formulations, for example, flour, sugar, baking powder, shortening, and water are used. In this case, one or more properties of the food product generally depend only on the proportions of the ingredients present in the mixture and not on the amount of the mixture. One ingredient (an independent variable) cannot vary without changing at least one of the other ingredients in the mixture, because all the ingredients will be part of a constant sum of 100%. In other words, the variables or the ratios of different ingredients in the recipe are dependent on each other. These phenomena do not meet the orthogonality requirement of a conventional factorial design. Therefore, to study and model the effects that different ingredient components in a mixture have on the food product properties of interest, the factorial experimental design (Chapter 3) is no longer suitable unless it is modified. For solving this kind of problems, another type of statistical experimental methodology—Mixture Experimental Design—is used. The effects of ingredient components (mixture variables) on food quality (response) are modeled differently from those effects based on the usual factorial experimental methodology.

As described above, the distinguishing feature of a mixture problem is that the independent or controllable factors represent proportionate amounts of the mixture rather than unrestrained amounts. These proportions are measured by volume, by weight, or by mole fraction. They are nonnegative numbers, and, if expressed as fractions of the mixture, they must add up to a unity, especially if the ingredients to be studied are the only ingredients comprising the mixture. For example, if a mixture consists of n components whose ratios in the mixture are $X_1, X_2, \ldots, X_i, \ldots, X_n$ respectively, then the sum of all these ratios would make a unit of which represents 100% of the ingredient [Equation (2.3)]. In other words, the ratio of any single component X_i is equal to unity minus the sum of all the other ratios [Equation (2.4)]. This means that any single component is defined by the other components, or the degree of freedom of a mixture system with n components equals $(n - 1)$.

$$\sum_{j=1}^{n} x_j = X_1 + X_2 + \cdots + X_n = 1.0 \qquad (2.3)$$

or

$$X_i = 1.0 - \sum_{j=1}^{i-1} x_j - \sum_{j=i+1}^{n} x_j \tag{2.4}$$

where $j = 1, 2, \ldots, n$.

It is not unusual in a formula that only some of the components would be selected for studying their effects on food product quality, resulting in the sum of ratio of these selected components as less than a unity. In this case, the variable proportions can be rewritten as scaled fractions so that the scaled fractions add up to unity. This is the fundamental restriction assigned to the proportions comprising the mixture experiment.

The geometric description or the coordinate system of a mixture system may not be familiar to most readers. The factor space containing the n components consists of all points on or inside the boundaries (e.g., vertices, sides or edges, and faces) of a regular $(n - 1)$ dimensional simplex. For a mixture with two components X_1 and X_2 $(n = 2)$ the simplex factor space is a straight line or axis (Figure 2.3). The two ends of the line ● represent the systems comprising pure X_1 or X_2, respectively; any blends or mixtures (X_1', X_2') of these two ingredients X_1 and X_2 are represented geometrically by a point on the line.

With three components, X_1, X_2, and X_3 $(n = 3)$ the simplex factor space is an equilateral triangle (Figure 2.4). The three vertices represent the 100% of components X_1, X_2, and X_3, respectively. The mixture represented by the sides or edges is only a two-component mixture, where the ratio of the component opposite to this line equals zero. In other words, a line is a two-component mixture system that is identical with the system illustrated in Figure 2.3. Every point inside the triangle simplex represents a mixture of X_1, X_2, and X_3. The coordinates of a point inside the simplex (black dot) will be (X_1', X_2', X_3'). The geometrical description of X_1', X_2', and X_3' is shown in Figure 2.4.

For a mixture system with four components X_1, X_2, X_3, and X_4 $(n = 4)$, the mixture simplex coordinate system is a tetrahedron and illustrated in Figure 2.5. Each of the four surfaces (ABC, ABD, ACD, and BCD) represents mixtures with only three components, namely, all the components except the one represented by the apex opposite to this surface. Each surface represents a simplified system of a three-component mixture and can be illustrated by a triangle as shown in Figure 2.4. The lines between two components (AB, AC, AD, BC, BD, and CD) represent mixtures with only two variables. Any point inside the tetrahedron is

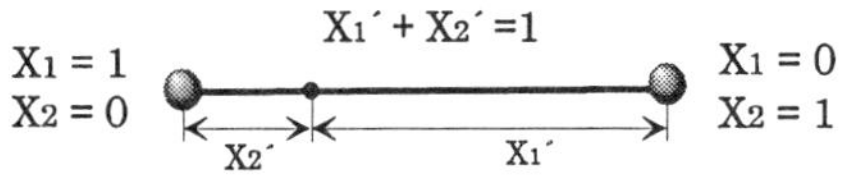

Figure 2.3 Geometric description of the simplex coordinate system for a mixture system with a mixture of two components X_1 and X_2.

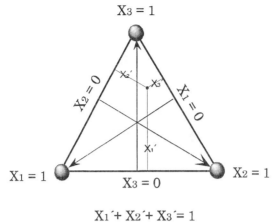

$$X_1' + X_2' + X_3' = 1$$

Figure 2.4 Geometric description of the simplex coordinate system for a system with a mixture of three components X_1, X_2, and X_3.

a mixture with four components X_1, X_2, X_3, and X_4, as illustrated by the point (X_1', X_2', X_3', X_4') shown in Figure 2.5. Geometrically, X_1', X_2', X_3', and X_4' is equal to the distance from the point to the faces opposite the corresponding apex.

For mixtures with more than four components ($n > 4$), the mixture simplex coordinate system is illustrated, but the principles applied are similar.

In a food formulation study, the mixture experimental approaches should be used to model the relationship between the component proportion (variables) and the food quality indices (responses) over the mixture region. The established models can then be used for recipe optimization. However, before

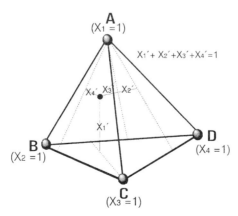

Figure 2.5 Geometric description of the simplex coordinate system for a system with a mixture of four components X_1, X_2, X_3, and X_4.

starting the modeling operation, a mixture experimental design, which is suitable for an efficient gathering of observations, should be selected. Based on these observations, a proper model is chosen.

In summary, the mixture experimental designs are unique because they take into consideration certain constraints. They are typically performed whenever the response is a function of the proportions of the variables and not their total quantities. The mixture experiments find wide applications in food formulation development. It is discussed in detail in Chapter 4.

2.5 DIFFERENCE OF MODELING FOR PROCESS AND RECIPE

Generally, all problems that appeared in food product design can be divided into mixture or process problems, with the latter having the dominant share. Sometimes a problem that seems to be a mixture problem is really a process problem and can only be solved with a corresponding factorial experimental method. As explained above, the difference between a process and a mixture study is quite distinct, and these studies need different statistical experimental techniques to deal with. In practice, it is not easy to distinguish a process problem from a mixture problem, when the food product design is only concerned with recipe or formulation development. To get a better understanding of the difference between them, a short description of performing a factorial experiment for solving a process problem and of running a mixture experiment is given:

(1) A factorial experiment: A factorial experiment studies the effect of some independent variables on food quality indices (response) through varying two or more of these independent variables, such as temperature, time, pressure, and pH value. A series of values or test levels of each factor is selected, and certain combinations of their levels are tested.
(2) A mixture experiment: An experiment in which the food quality indices (response) are assumed to depend only on the relative proportions of the ingredient components present in the mixture and not on the amount of the mixture is considered a mixture experiment. In such an experiment, if the total amount of the mixture is held constant, the value of the response changes when changes are made in the relative proportions of the ingredients.

The development of bakery powder is described as a practical example that will help in understanding the difference between a factorial and a mixture experiment.

A premixed baking powder for biscuit making consists of wheat flour F and three different chemical compounds A, B, and C, which would be tested in biscuit making according to a standard bakery experiment. The flour is used as a diluting medium, whereas A, B, and C will be effective at different baking

TABLE 2.1. Formulations Used for Strategy I.

Chemicals (g)				Flour (g) F	Total (g)	Chemicals (%)				Flour (%) F	Total (%)	Ratio of Chemical:F
A	B	C	Total			A	B	C	Total			
40	10	10	60	3940	4000	1.00	0.25	0.25	1.5	98.5	100	3:197
30	15	15	60	3940	4000	0.75	0.375	0.375	1.5	98.5	100	3:197
30	20	10	60	3940	4000	0.75	0.50	0.25	1.5	98.5	100	3:197
20	20	20	60	3940	4000	0.50	0.50	0.50	1.5	98.5	100	3:197

temperatures or baking phases. To develop an optimal baking powder formulation from F, A, B, and C, the effect of various formulations are tested. Three different statistical experimental approaches are applied.

2.5.1 STRATEGY I

Wheat flour F 3940 g is mixed with 60 g of A, B, and C, yielding a ratio of chemicals to flour of 3:197 in all formulations. All mixtures are produced in total amounts of 4000 g. In all tests, the amount of flour F in the baking powder is fixed, and the amounts of the three chemicals A, B, and C are varied (Table 2.1).

It is clear that what is of interest here is to find out how different proportions or percentages of A, B, and C influence the biscuit quality when they are mixed with wheat flour in a 3:197 ratio. This is a three-component mixture experiment in which the amount of each chemical A, B, and C is varied, but their total amount is kept constant at 60 g (A + B + C = 60 g), and the ratio of chemicals to flour is fixed at 3:197 in all combinations. In such an experiment design, the effect of A, B, and C on the biscuit quality can be determined.

2.5.2 STRATEGY II

The wheat flour amount is fixed at 3940 g, but that of the active ingredients is varied from 30 to 60 g with the ratio of A:B:C fixed at 3:2:1. The combinations to be tested are listed in Table 2.2.

TABLE 2.2. Formulations Used for Strategy II.

Chemicals (g)				Flour (g) F	Total (g)	Chemicals (%)				Flour (%) F	Total (%)	Ratio of Chemical:F
A	B	C	Total			A	B	C	Total			
30	20	10	60	3940	4000	0.750	0.500	0.250	1.500	98.500	100	0.0152
25	16.7	8.3	50	3940	3990	0.627	0.418	0.209	1.254	98.746	100	0.0127
20	13.3	6.7	40	3940	3980	0.502	0.334	0.167	1.004	98.996	100	0.0101
15	10	5	30	3940	3970	0.378	0.252	0.126	0.756	99.244	100	0.0076

TABLE 2.3. Formulations Used for Strategy III.

Chemicals (g)				Flour (g) F	Total (g)	Chemicals (%)				Flour (%) F	Total (%)	Ratio of Chemical:F
A	B	C	Total			A	B	C	Total			
30	20	10	60	3960	4020	0.746	0.497	0.249	1.5	98.50	100	0.01523
30	20	10	60	3940	4000	0.750	0.500	0.250	1.5	98.50	100	0.01523
15	10	5	40	3960	4000	0.375	0.250	0.125	1.0	99.00	100	0.01010
15	10	5	40	3940	3980	0.377	0.254	0.127	1.0	99.00	100	0.01010

It is a single-factor experiment with four test levels of baking powder. Actually, only the ratios of chemicals to flour are changed in the study. In this way the effect of changing the chemicals:flour ratio or of changing the amount of chemicals while holding the amount of flour constant can be measured. In this experimental design the effect of the ratio of chemicals to flour would be examined. Note that if the percentages of A, B, and C are varied in addition as in the first experiment, this would then constitute a four-component, mixture-amount experiment.

2.5.3 STRATEGY III

Two levels of wheat flour 3960 g and 3940 g and two levels of baking powder 60 g and 40 g are selected to be tested. In all trials the percentages of A, B, and C are fixed at 3:2:1. The formulations are as in Table 2.3. This is obviously a factorial experiment in which we are interested in measuring how the biscuit quality will be influenced by changing the level of flour and chemicals.

CHAPTER 3

Food Process Modeling and Optimization

3.1 BASIC CONCEPTS OF FOOD PROCESS DESIGN

A S described in Chapters 1 and 2, the mathematical modeling and optimization technique is an important and effective approach to solve problems encountered in food product design. These problems can be divided into two categories: process and recipe modeling. This chapter introduces the mathematical modeling and optimization technique in a process system.

To successfully model a process during the food product design, familiarity of the personnel in charge of the product development with this process is required. They also should be acquainted with the procedures of the mathematical modeling technique. Only qualified personnel would ensure correct problem identification, solid data gathering, and data analysis, which lead to reliable model building and process optimization. On the basis of the model built, the food researcher can identify the most important or potentially important factors or variables that directly or indirectly affect the targeted product quality indices.

Mathematical modeling is based on the assumption that the relationship between the variables and the food quality indices in a food production process can be described in terms of a mathematical equation. When some experiments were specifically designed and served as support points, an empirical model could be established to summarize the process studied in the form of a mathematical function. This function can improve the understanding of the process and the effect of various variables on the desired food quality attributes (objectives). It also can reveal critical points or regions that are not obvious before modeling and supply information about variables for optimum food production. Additionally, it provides an overall view and guide for the food product researcher and serves as a useful tool for food process or product design. Of course, only a precise model is valid and can be used to solve problems such as food quality prediction, interpolation, and optimization.

3.2 IDENTIFICATION OF MODELING SYSTEM

3.2.1 OBJECTIVE IDENTIFICATION AND DEFINITION

As discussed in Chapter 1, before modeling a food process system for product design, the system must be first analyzed, and responses of interest and factors affecting the responses have to be identified and defined. Thus, an intimate understanding of the food materials, machines, processes and the marketing objectives is essential for doing food product design. An overall and clear picture of the food product quality is the initial requirement for a successful product development project. Identification of the independent variables in a process, which may be different for different food quality objectives, is the second essential requirement for successful modeling.

To identify the responses means to find out the most important product quality indices or criteria that are essential and significant for the food to be developed. In practice, it is not difficult to reveal and define these quality objectives. Normally, reliable and repeatable sensory, physical, chemical, microbiological, and even rheological methods must be selected for quantitative measurement of these quality indices of interest. However, specific and adequate measurement methods must be applied for food products of different categories.

If the food developed is a product and goes directly to the consumer market, then its sensory quality is particularly important. In most cases, the sensory evaluation seems to be the easiest. Actually, it is the most difficult objective to be determined because each individual consumer has his own sensory preferences and evaluation standards. A complete sensory evaluation used for food product design can be performed by a small panel of 5–20 panelists according to a predesigned tabular schema. Thus, the schema must be prepared with great care so that all the characteristics important for the acceptance of this specific food are included. The environmental conditions (e.g., temperature, humidity, time, and the ways a consumer uses the product) affecting consumption of the product should be considered as well. A schema, as an example for sensory evaluation of instant powder food, is given in the Appendix. Furthermore, the panelists should include people who are familiar with the specific food product or similar products. It must be mentioned that each kind of food requires a specific panel for its sensory evaluation. For example, a qualified wine sensory evaluator is not suitable to make a sensory evaluation of snack foods, if he is not trained for this.

The physical, chemical, and nutritional values of a food product are normal quality objectives. They can be measured or calculated reliably with negligible error with standard laboratory procedures. Replications should be used to obtain a reliable set of physical and chemical data.

It must be always kept in mind by the food product developer that an incorrect specific description or insufficient definition of the product quality indices would result in an inaccurate optimum food product. Some of the common

TABLE 3.1. Some General Food Objectives and Determining Methods.

Categories	Quality Indices	Measurement Methods
Sensory criteria	Size and appearance	
	Color	Sensory evaluation
	Texture	Physical and/or rheological measurements
	Taste	
	Flavor	
Nutritional value	Energy	Chemical or calculation methods
	Protein content	Chemical, biological, or calculation methods
	Protein nutritional quality	Biological, chemical, or calculation methods
	Lipid content	Chemical or calculation methods
	Carbohydrate content	Chemical, enzymatic, or calculation methods
	Soluble cellulose content	Chemical methods
	Insoluble cellulose content	Chemical and physical methods
	Moisture content	Chemical or physical methods
	Digestibility	Enzymatic or animal-feeding methods
	Antinutritional factors	Enzymatic or animal-feeding methods
	Vitamins	Enzymatic or chemical methods
	Mineral elements	Chemical and physical methods
	Biological active factors	Medical or biological methods
Physical and rheological properties	Shear force	Physical methods
	Viscosity	Viscometer
Functional properties	Specific volume	Physical methods
	Water adsorption index	Physical methods
	Water soluble index	Physical methods
Storage property	Shelf life	Sensory, chemical, and microbiological test
Cost	Ingredient and manufacturing cost	Normal calculation, linear programming

food quality attributes and the corresponding determination methods for food products of all categories are listed in Table 3.1. They may be considered as stages in food quality (response) identification and definition.

In the area of food development, many of the product quality indices, such as taste and texture, might be qualitative variables. However, they must be transformed into corresponding numerical variables to enable mathematical modeling. For a certain food product there is usually more than one product quality index to be considered and optimized. Each quality index must be modeled on the basis of the same experimental design (real or coded) and on the corresponding real objective values. For example, if five food quality indices are selected for the target product, then five individual models must be built for each of them. The optimization can then be performed according to all of these five models.

3.2.2 FACTOR IDENTIFICATION AND SCREENING

It is usually not easy to identify independent variables critical or related to the food product quality and to account for most of the variations in the product as they occur in the process system. Nevertheless, to analyze the process of interest, a block flowchart is helpful for a better understanding and an overview of the variables in the whole system. All the subprocesses should also be analyzed. The suspected influencing factors should be listed and their relationships illustrated. If there seem to be too many variables involved (more than 6–8), as is the case in most new food product development, then a screening procedure should be used to identify those that are important to the responses of interest. Those significant variables are then selected for further optimization.

To solve the screening problem, two statistical approaches, the Plackett-Burman design and the saturated fractional factorial design, have proved to be suitable. In these approaches, some specifically designed preliminary experiments are conducted. They enable the food researcher to roughly estimate the effect of each factor and to select the most significant and important variables from all suspected ones with minimum experimental efforts. Those significant variables selected can then be used for further experimentation. During the screening stage, attention is concentrated on gathering qualitative information about which specific independent variables significantly affect the different specific product quality indices.

However, if some important variables are not identified and tested in a further experimental plan, the subsequent statistical product development procedure cannot involve them in the statistical model. This would lead to inexact or wrong optimization results at the last stage of food product design.

3.2.2.1 Plackett-Burman Design

The Plackett-Burman design is a well-known screening approach that was developed by Plackett and Burman in 1946. It allows the developer to test the largest number of factor main effects with the least number of trials. Actually, Plackett-Burman designs are fractionalized full factorial designs in which the number of runs is a multiple of four. We can use all available factors in a Plackett-Burman design. However, sometimes we want to generate a saturated design for one more factor than we are expecting to test. In an extreme case, we use $(n + 1)$. Usually, we use $(n + 4)$ trials for qualitatively examining the significance of n factors, without considering possible interactions. This will allow us to estimate the random experimental error variability and test for the statistical significance of the estimated parameters. Sometimes the center points and replications are added to the design to test the curvature tendency and to increase the available degree of freedom for error estimation. Foldover is another method used to increase the resolution of a Plackett-Burman design.

Several standard Plackett-Burman designs are involved, and each tested at two levels without replication is listed in the Appendix. The following example illustrates the principles of screening techniques. The sponge-and-dough procedure is a popular bread-baking process. In this procedure, part of the flour, water, and yeast are mixed to form a loose dough (sponge). The sponge is then fermented, combined with the rest of the formula ingredients, and mixed into developed dough. After being mixed, the dough is given an intermediate proofing so that it can relax. Finally, it is divided, molded, proofed, and baked.

The following are many factors in the sponge stage that may have potential effects on the specific bread volume (ml/100 g flour):

(1) proofing temperature of sponge [SpgTemp (°C)]
(2) yeast amount in sponge (percentage of the total flour amount)
(3) type of flour
(4) flour part used in sponge (percentage of the total flour amount)
(5) proofing time of sponge (h)
(6) salt content (percentage of the total flour amount)
(7) sponge dough yield (DoughYD) (g/100 g flour)
(8) water content of the dough

The normal varying ranges of these variables are shown in Table 3.2. In the example, a Plackett-Burman design with 12 trials was used to identify the most significant factors from these 8 for further optimization study. Table 3.3 shows the design and the trial result. These data are analyzed with the Experimental Design module of the computer program STATISTICA. The analysis of variance (ANOVA) technique is used to investigate the significance level of each variable effect, and the ANOVA result is shown in Figure 3.1. In general, the ANOVA can be performed by using most of the statistical software listed in Chapter 1.

TABLE 3.2. Influencing Factors and Their Levels for Test.

	Coded Lower Level	Coded Upper Level	Real Lower Level	Real Upper Level
SpgTemp	−1	+1	20	30
Yeast	−1	+1	0%	2%
FlourType	−1	+1	Sort A	Sort B
FlourPart	−1	+1	20	60
Time	−1	+1	8	15
Salt	−1	+1	0	1.5
DoughYD	−1	+1	150	250
Water	−1	+1	43%	47%

TABLE 3.3. Plackett-Burman Design and Trial Results.

Trial No.	Spg Temp	Yeast	Flour Type	Flour Part	Time	Salt	Dough YD	Water	I	J	K	Volume
1	1	−1	1	−1	−1	−1	1	1	1	−1	1	672
2	1	1	−1	1	−1	−1	−1	1	1	1	−1	723
3	−1	1	1	−1	1	−1	−1	−1	1	1	1	642
4	1	−1	1	1	−1	1	−1	−1	−1	1	1	666
5	1	1	−1	1	1	−1	1	−1	−1	−1	1	753
6	1	1	1	−1	1	1	−1	1	−1	−1	−1	693
7	−1	1	1	1	−1	1	1	−1	1	−1	−1	687
8	−1	−1	1	1	1	−1	1	1	−1	1	−1	639
9	−1	−1	−1	1	1	1	−1	1	1	−1	1	623
10	1	−1	−1	−1	1	1	1	−1	1	1	−1	715
11	−1	1	−1	−1	−1	1	1	1	−1	1	1	722
12	−1	−1	−1	−1	−1	−1	−1	−1	−1	−1	−1	645

Check the significance level in Figure 3.1, and find that the effect of the proofing temperature (SpgTemp), yeast amount, the type of flour (FlourType), and the dough yield (DoughYD) are significant (marked with **). These significant variables should be chosen for further optimization study.

3.2.2.2 Saturated Fractional Factorial Design

The saturated fractional factorial design is another frequently used statistical variable screening approach. It allows a food product developer to test only a fraction of the factor combinations in a full factorial design and still be able to obtain the most important system information. The trial number depends on the degree of the design saturation. However, one or two trials at the center point are usually included in the design to check the curvature properties of the

ANOVA: Response: VOLUME; R-sqr=.9777; Adj:.9183;
8 Factor Screening Design; MS Residual=135.222

	SS	df	MS	F	p	Signif.
SpgTemp	5808.000	1	5808.000	42.952	0.007	**
Yeast	5633.333	1	5633.333	41.659	0.008	**
FlourType	2760.333	1	2760.333	20.413	0.020	**
FlourPart	0.333	1	0.333	0.002	0.963	
Time	208.333	1	208.333	1.541	0.303	
Salt	85.333	1	85.333	0.631	0.485	
DoughYD	3201.333	1	3201.333	23.675	0.017	**
Water	108.000	1	108.000	0.799	0.437	
Error	405.667	3	135.222			
Total SS	18210.670	11				

Figure 3.1 Analysis of the trial results of the screen design with help of the software STATISTICA.

system. Similar to the Plackett-Burman design, a saturated fractional factorial design can only estimate the linear effect of the factors but cannot estimate any interactions between them. This means that the model selected for the screening stage of product development contains terms for the individual factors but may not include interaction effects between factors. However, such designs are usually only used to check the significance of a number of possible variables and enable the selection of significant ones for further precise modeling and optimization. The analysis of the result is similar to that of a Plackett-Burman design. For more information about saturated fractional factorial designs read Section 3.4.3.2 and related reference books listed in the Appendix.

3.3 SELECTION OF SUITABLE MODELS

After selection of the food quality objectives (response) of interest and the significant independent variables for further tests in the process system, the next step of statistical food product design is to establish an experimental design. The trials in the plan would support the building of a mathematical equation, which shows the dependence of the objective food quality indices (responses) Z on the influencing factors (independent variables) X_i [Equation (3.1)] with sufficiently high precision:

$$Z = f(X_i) \tag{3.1}$$

It is difficult to select a suitable mathematical equation to describe a definite food process system. In principle, different empirical models or mathematical equations can be built up by fitting them mathematically to the same set of organized experimental data. Figure 3.2 shows the possibility of fitting a set of experimental data to three different models that are plotted as lines or curves in the figure. Two criteria to consider for choosing a usable and precise model from the many possible equations are as follows:

- the model with the highest precision for accurate application
- the model with the simplest form for easy application

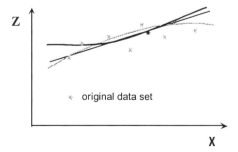

Figure 3.2 Three possible curve fits based on a same data set.

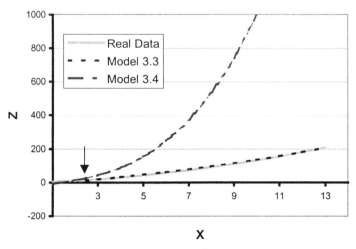

Figure 3.3 A same set of data can be fitted by different models with different accuracy.

Actually, the different models used to fit the same set of data differ from each other in fitting accuracy. For example, a true relationship (real model) which is described by Equation (3.2) can be fitted by two other models [Equations (3.3) and (3.4)] in terms of regression:

$$Z = 1 + 3 \times X + X^2 \tag{3.2}$$

$$Z = -19.04 + 12.39 \times X + 0.0296 \times X^3 \tag{3.3}$$

$$Z = -19.04 + 12.39 \times X + 0.0296 \times X^2 + 0.8856 \times X^3 \tag{3.4}$$

For a better understanding of the difference between these three models, they are plotted in Figure 3.3. The data from the real model [Equation (3.2)] can be predicted almost perfectly by Model 3.3. However, Model 3.4 approaches the real model only in the small X values ranging up to three and swiftly deviates far off with increasing X values. In practice, this means that Model 3.4 can only be used in the small X value range to predict the data of the real model with good accuracy. The negligible difference between the curves of the real model [Equation (3.2)] and Model 3.3 indicates that Equation 3.3 is an acceptable estimate for Model 3.2. The slight deviation in the figure indicates, however, that a mathematical regression can rarely generate a model that describes the original data with 100% accuracy.

In food product design, any modeling must correspond to reality, and this requires some actual experience. It needs the observed experimental results of experiments to guide the form of the model and to ensure that the model describes the actual situation. Polynomials have been used extensively in empirical modeling of chemical, biological, and food research systems. They provide a simple curvilinear relationship between a large number of variables, possess a clearly defined optimum, and use simple computational algorithms by using the least square method for estimation of the coefficients β in the model. Furthermore, a polynomial can readily be used through appropriate mathematical transformations of the independent or dependent variables.

The simplest polynomial model is a linear equation [Equation (3.5)]. It is constructed of only additive terms, each containing only one multiplicand parameter. Unfortunately, this model does not achieve sufficient precision because of the frequent occurrence of interactions between variables in a process. It is believed that low-degree polynomials, such as a first-degree polynomial with interaction terms or a quadratic polynomial, are the most appropriate models to adequately describe a food process. Therefore, most of the statistical experimental methodology is based on supporting the building of a Taylor expansion equation up to second degree, which would meet the requirements of food product design. A first-degree Taylor expansion equation [Equation (3.6)] can be used to model a simple food process, because fewer experiments and expenses are required to build it. A Taylor expansion equation of the second degree [Equation (3.7)] may be used if the first-degree expansion equation cannot achieve an adequate descriptive precision of a complicated system. In this chapter, some typical experimental designs are introduced to support the building of Taylor expansion equations of the first and second degrees.

$$Z = \beta_0 + \sum_{j=1}^{n} \beta_j \cdot X_j \tag{3.5}$$

$$Z = \beta_0 + \sum_{j=1}^{n} \beta_j \cdot X_j + \sum_{j<k=2}^{n} \beta_{jk} \cdot X_j \cdot X_k \tag{3.6}$$

$$Z = \beta_0 + \sum_{j=1}^{n} \beta_j \cdot X_j + \sum_{j<k=2}^{n} \beta_{jk} \cdot X_j \cdot X_k + \sum_{j=1}^{n} \beta_{ij} \cdot X_j^2 \tag{3.7}$$

where β_0 is a constant, β_j, β_{jj}, and β_{jk} are coefficients, and

$$\sum_{j=1}^{n} \beta_j \cdot X_j \text{ are linear terms.}$$

$$\sum_{j<k=2}^{n} \beta_{jk} \cdot X_j \cdot X_k \text{ are interaction terms.}$$

$$\sum_{j=1}^{n} \beta_{jj} \cdot X_j^2 \text{ are quadratic terms.}$$

A full quasi-first-degree polynomial model [Equation (3.6)] includes terms of linear and interaction effects between factors. The interaction effects represented by terms $X_j \cdot X_k$ indicate physically that the effect of X_j on the response depends on the level of X_k. The interaction effects make the response plane surface generated according to the model more or less bent. Similarly, a full second-degree polynomial model [Equation (3.7)] includes terms of linear, interaction, and curvature effects. Curvature effects, represented by terms such as X_j^2, produce parabolic shapes when the model is plotted. This effect occurs when two different levels of the same factor produce similar values of response and higher or lower responses at middle levels.

It is much easier to handle a first-degree polynomial equation than the higher degree ones. A low-degree polynomial contains fewer terms and requires fewer trials as support points (trials) to estimate the coefficients in the equation. In practice, the choice of the model form depends on the complexity of the practical food system. Some complicated systems require a polynomial of the third degree. Suitable transformation of the experimental data may be considered if the fitting of the data with a normal model of the first or second degree turned out to be inadequate. Fortunately, quadratic models have proved adequate in most cases of food product design.

The model form depends also on the range of experimental space of the process system. This experimental space is built through limitations of the extreme levels of all the independent variables selected for test. Normally, a narrow space can be described with a linear equation with or without interaction terms, although the real relationship is quadratic. Otherwise, a quadratic model should be adopted to model the system. One of many advantages of the statistical product design is that the experiments can be designed in such a manner that a model of first degree, with or without interaction terms, or of second degree can be supported. This model can describe the system with adequate precision.

3.4 DESIGN OF STATISTICAL EXPERIMENTAL PLANS

3.4.1 SELECTION OF TEST LEVELS

Once the responses are defined, the significant variables determined, and the appropriate approximating polynomial selected according to the process complexity, an appropriate experimental design can be formulated. This experimental design would summarize the essential process information in a mathematical model. Design of experiment is a critical step for a successful food product study because only well-constructed experiments are easy to analyze and provide reliable information. The property of the polynomial used to estimate the response function depends largely on the pattern of the trials in the experimental design. Poorly designed experimentation requires extensive and complicated analysis of data, which might not support significant modeling. Furthermore, food experimentation is often time-consuming and expensive. In many cases, such as pilot-plant runs and factory trials, the size of the experiment becomes extremely important.

Although covering the range of factor levels specified in the experiment, a statistical experimental design emphasizes those trials on the borders and midpoints of the ranges to be researched, thereby decreasing the total number of trials with minimal loss of information. The basic properties of a statistical experimental design should be:

- The effects of one or more variables should be turned off by means of keeping it/them at constant level(s) in some trials. In this way its/their effect(s) on the response can be shut off and the influences of the remaining variables, which still change in these trials, can be determined.
- The test levels of each variable in the trials must be selected in such a way that no significant correlation between the independent variables exists in the whole experimental design.
- The experimental levels of every variable are selected in such a manner that a near square research space will be defined for well-balanced supporting.
- It is necessary to arrange as many experimental points as possible for each level of a variable.
- The number of repeated trials must be high enough to control random errors. All trials in the experimental design should be randomized to minimize unknown errors.
- The design must supply more degrees of freedom than the model needs, i.e., the number of trials in an experimental design must be greater than the number of terms (except the constant term) in the model.

Successful modeling of the process system is one of the essential steps in food product development. A well-designed statistical experimental plan supports the reliable study on the single and combined effects of independent variables on some specific quality criteria of the food product. In addition, this systematic approach allows food scientists and technicians to obtain the greatest amount of meaningful information from the fewest number of experiments at the lowest cost.

As discussed above, generating a statistical experimental plan for food product design is to define the level ranges or test levels of each process variable selected in a meaningful way. The range of one variable is defined by two boundary values, which consists of a low and a high level. All the low and high levels of each variable then build a room or space in which the region of interest for food researchers should be located. The model established in the following stage is in principle only valid and effective within this range.

The experimental levels selected for a test must make sense. Hence, experience with the food process is usually required. For example, in developing bread products, selected dough fermentation temperature should not vary from 5 to 50°C, because such temperatures are not significant for dough fermentation. A reasonable temperature range might be from 20 to 40°C.

In general, the spectrum of test levels of one independent variable should be quite broad so that a wide range can be researched and more information can be achieved. However, if the level span is too large, the model built might not be significant (not exact enough); and if the level interval of one variable is too narrow, this variable would lose its significance in the model.

In food research, some independent variables, such as sorts of flour, types of enzyme, or brands of machinery, are qualitative or category factors. This means that they cannot vary continuously in the whole experimental range but can only be set at definite levels. These kinds of variables should be carefully transformed into quantitative ones to allow the modeling of their effects in the system. For example, three sorts of flour can be defined as three levels, that is, 100, 101, and 102, or -1, 0, and 1. However, it must be kept in mind that level 101 is not larger than 100 or less than 102. A category variable cannot be set at any level other than these three numerical ones during regression and optimization operations.

The model form itself could depend on the level spans of each variable. As illustrated in Figure 3.4, even a curved line can be described in the form of a linear line if a narrow range is selected for research (range c). Selecting a reasonable test range is accomplished on the basis of experience and results from pretrials. In practice, a small interval of test levels would make sense only if it is ensured that the optimum range is contained in this experimental space. If the test levels are under the control of the food researcher, they should be set fairly broad (of range a in Figure 3.4). Sometimes a second experiment with a narrower range should be conducted to yield a more accurate representation of the food system.

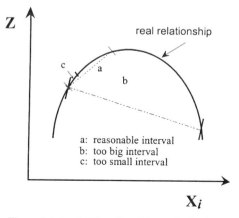

Figure 3.4 Possibilities of level interval selections.

3.4.2 EXPERIMENTAL DESIGN

Based on the model form, as well as on the selection of the number of variables and their test levels, an experimental design for trial can be constructed. It should be drafted according to a suitable standard coded statistical experimental design, whose construction is based on statistical principles and fulfills the requirements listed in Section 3.4.1. A standard experimental design is presented in a neutral form with codes, in which "-1, 0, and $+1$" or "low, middle, and high" refer to the variable levels. Some typical and frequently used standard experimental designs are listed in the Appendix. Each variable X_i is arranged in one column in a coded design.

A real experimental plan is easily established through rewriting the coded experimental design with the coded lower and upper levels being replaced with the real lower and upper levels of each variable. During the rewriting, the level combination of each variable in each single trial (in a row) must be kept unchanged. A formula [Equation (3.8)] can be used to rewrite a coded standard experimental plan into a real plan. This formula, which may be used in a computer program, is also useful for interpreting the optimization results based on a coded mathematical model into actual values. Obviously, this transformation can be reversed as described in Equation (3.9).

$$X_{i.\text{real}} = \frac{X_{\text{real.}+1} + X_{\text{real.}-1}}{2} + \frac{X_{\text{real.}+1} - X_{\text{real.}-1}}{2} \times X_{i.\text{code}} \qquad (3.8)$$

$$X_{i.\text{code}} = \frac{X_{i.\text{real}} - \dfrac{X_{\text{real.}+1} + X_{\text{real.}-1}}{2}}{\dfrac{X_{\text{real.}+1} - X_{\text{real.}-1}}{2}} \qquad (3.9)$$

where $X_{i.\text{real}}$ is the real value of any coded value $X_{i.\text{code}}$, $X_{\text{real.}+1}$, and $X_{\text{real.}-1}$ are the upper and lower real levels corresponding to coded levels $+1$ and -1.

3.4.3 EXPERIMENTAL DESIGN FOR MODELS OF THE FIRST DEGREE

3.4.3.1 2^n Factorial Experimental Design

3.4.3.1.1 Introduction

A 2^n factorial design is one of the most often used experimental plans in technical research. It is the most popular design among full factorial designs originally introduced by Fisher and is commonly used for screening and optimization in different research areas. In this design, the term *factorial* means that all factors are varied simultaneously over several test levels. The trial number of a full factorial experimental design depends on the number of independent variables included and their test levels. This design enables the food product developer to determine the main linear effects of each factor and the interactions between factors. However, such a design can be complex and requires many trials if there are more than four factors and more than three levels to be investigated.

The usual notation for describing a full factorial design is m^n, where n represents the number of factors and m the number of levels of each factor. The value of m^n equals the total trial number in the plan. Among full factorial designs, the 2^n and 3^n factorial designs are often used in food product design. The former design supports building models of the first degree with/without interaction terms [Equations (3.5) and (3.6)], whereas the latter supports the building of linear and full quadratic models [Equation (3.7)].

In a 2^n factorial design, n means the variable number chosen for investigation, and 2 is the two test levels (high and low levels) of each independent variable. 2^n equals the trial number in the design. A 2^n experimental design supplies 2^n degrees of freedom, in which one is fixed for determining the total average value β_0 (constant term) in the model. The remaining $(2^n - 1)$ degrees of freedom then allow $(2^n - 1)$ independent comparisons between the data of all those trials by which the interactions between the independent variables can be estimated and calculated. In a practical situation, if the variable number is fewer than 4, the trials in a 2^n experimental design should be repeated wholly or partly to get solid and reliable modeling results. In this case, the remaining degrees of freedom, which are used for error estimation, are too small.

3.4.3.1.2 General Principles of 2^n Factorial Designs

As discussed in Section 3.4.3.1.1, a 2^n factorial experimental design enables the food researcher to estimate the linear effects of all independent variables and their interactions on one or more objectives through mathematical modeling. For example, to model a system with two variables A and B, the model form as described by Equation (3.10) with an interaction term is used. To do this, only two test levels of variables A and B are selected and included in the 2^2 factorial experimental design. These two levels are normally boundary values of the research range. More than one response can be modeled simultaneously on the basis of the same experimental design using their corresponding response data and model.

$$
Z = \underset{\substack{\Updownarrow \\ \text{Total} \\ \text{average}}}{\beta_0} + \underset{\substack{\Updownarrow \\ \text{Linear effect} \\ \text{of factor } A}}{\beta_1 \cdot A} + \underset{\substack{\Updownarrow \\ \text{Linear effect} \\ \text{of factor } B}}{\beta_2 \cdot B} + \underset{\substack{\Updownarrow \\ \text{Interaction} \\ \text{effects } AB}}{\beta_{12} \cdot A \cdot B} \tag{3.10}
$$

A 2^n factorial experimental design consists of all possible combinations of the two test levels of all n factors. As an example a coded 2^3 factorial

TABLE 3.4. Coded 2^3 Factorial Design.

No.	X_1	X_2	X_3	Symbol
1	−1	−1	−1	
2	−1	−1	+1	
3	−1	+1	−1	●
4	−1	+1	+1	
5	+1	−1	−1	
6	+1	−1	+1	
7	+1	+1	−1	
8	+1	+1	+1	

experimental design is listed in Table 3.4. Factorial designs for two to six independent variables are listed in the appendix.

Each of the independent variables X_1, X_2, and X_3 would be arranged in a column in a table, with each column being used for one variable only. The "+1" and "−1" in the design are coded test levels corresponding to the real (actual) high and low test levels of X_1, X_2, and X_3. This coded plan can be transformed into an actual experimental design by replacing the +1 and −1 with the values of the actual high and low test levels. The selection of a column in the coded plan for one factor is not restricted. Any column can be chosen for one independent variable. All columns in the table are equivalent with each other in a full 2^n factorial design. The 2^3 design is described graphically in Figure 3.5. ● represents trials, and all trials build a cubic space, which is the researched space.

Almost all the computer programs specific for experimental designs can generate these kinds of designs. Coded 2^n factorial designs can be calculated according to the formula below [Equation (3.11)]. With use of the formula, a small computer program can be easily written to generate any coded 2^n full factorial experimental designs. It is not difficult to find that the level of each

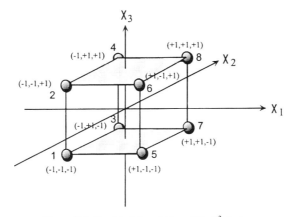

Figure 3.5 Graphical description of the 2^3 design.

variable varies in a cyclic way in a 2^n factorial design. Being aware of this, one can design a 2^n factorial experimental design directly.

$$X_{pq} = [-1]^{\text{INTEG}\left[\frac{q}{2^{(p-1)}}\right]}$$ (3.11)

where

p: $1, 2, \ldots, n$ is the row number in the coded 2^n factorial design

q: $1, 2, \ldots, 2^n$ is the column number (trial number) in the coded 2^n factorial design

INTEG yields the largest rounded whole number that is smaller or equal to the argument

The 2^n factorial design is a simple and effective statistical tool used to solve modeling problems in food product design. A 2^n design is based on the hypothesis that the relationships between the dependent and the independent variables in the researched area are linear. It also has been proved that most food systems can be well modeled with a first-degree model [Equation (3.6)] based on a 2^n factorial design. However, a 2^n factorial experimental design would not support the estimate of nonlinear (often quadratic) properties between one factor and the responses within the two test levels. To determine the quadratic effects statistical design of the second order must be applied.

3.4.3.1.3 A Practical Example

A cooking extruder is used to inactivate the trypsin inhibitor (TI) in soybean flour, in which the effect of the three most important operating factors (independent variables) are studied:

- product temperature T (°C)
- mass moisture content M (%)
- screw rotation speed n (min^{-1})

A 2^3 factorial design was used to investigate the effect of these three variables on the inactivation percentage degree of the trypsin inhibitor D_{TI} (%) through extrusion cooking. The coded and real test levels of T, M, and n are listed in Table 3.5.

TABLE 3.4. Real and Coded Test Levels of T, M, and n.

	T	M	n
Coded lower level	−1	−1	−1
Coded upper level	+1	+1	+1
Real lower level	130	18	120
Real upper level	170	30	160

TABLE 3.5. Real Experimental Design.

No.	T	M	n
1	130	18	120
2	130	18	160
3	130	30	120
4	130	30	160
5	170	18	120
6	170	18	160
7	170	30	120
8	170	30	160

According to the coded experimental plan in Table 3.4, a real experimental design for trial can be constructed (Table 3.6) by replacing -1 with 130°C for T, 18% for M, and 120 min^{-1} for n; $+1$ with 170°C for T, 30% for M, and 160 min^{-1} for n, respectively.

3.4.3.2 2^{n-p} Fractional Factorial Designs

A full factorial design would work fine, except that the number of necessary runs in the experiment will increase geometrically. Because each run may require time-consuming and costly setting and resetting of variable levels, it is often not feasible to require that many different trials for the experiment. Besides the interaction effects between two independent variables, the full 2^n factorial designs also permit estimation of interactions between more than two factors. These kinds of interaction effects are usually not significant and have no physical meaning. The degree of freedom used in estimating them are therefore assigned for that of the experimental variance. However, this degree of freedom can also be used to estimate the effects of additional p factors. In other words, the highest order interactions were successively used to generate new factors. The basic idea to set up a 2^{n-p} fractional factorial design. Fractional means that only a fraction of the factor combinations in a full factorial design would be chosen for experimentation. 2^{n-p} designs allow food researchers to obtain important information with an economical number of trials and thus make themselves an effective tool at the stage of identifying and screening of significant independent variables.

Box and Hunter have developed several rules and algorithms for the construction of 2^{n-p} fractional factorial designs. In principle, the columns in the experimental design for estimation of the highest interaction should be used for testing new variables first. For example, if a 2^{4-1} design is used with eight trials, then besides the three variables X_1, X_2, and X_3 the fourth variable X_4 can be set in the column generated by $X_1 \cdot X_2 \cdot X_3$. The following example shows this principle more concretely.

To check the effect of five independent variables X_1, X_2, X_3, X_4, and X_5 on a target parameter Y, normally a 2^5 factorial plan with $2^5 = 32$ trials was required. If it is ensured that one interaction effect of three variables, say X_1 X_2 X_3, is not significant, then the capability to estimate the effect of X_1 X_2 X_3 through a 2^4 design can be used to estimate the effect of the fifth variable X_5. The number of needed runs of a 2^{5-1} design is half of a 2^5 design. The column for estimating the interactive effect of the highest level X_1 X_2 X_3 X_4, which is usually of no practical meaning, can also be used to assign a new variable X_5. During construction of the 2^{5-1} experimental design, a normal 2^4 factorial design was modified. A new column assigned to X_5 was built by multiplying the three level values in these columns of X_1, X_2, and X_3 (Table 3.7). This calculation can be easily performed in a spreadsheet of Ms-Excel®. If X_1, X_2, X_3, and X_4 are assigned in columns A, B, C, and D, respectively, the column for X_5 can be calculated by using the formula " $= A1 \cdot B1 \cdot C1 \cdot D1$." A 2^{n-1} fractional factorial design, also known as $\frac{1}{2} 2^n$ design, still supports the estimate of two-factor interactions $X_1 \cdot X_2$, $X_1 \cdot X_3$, $X_1 \cdot X_4$, $X_2 \cdot X_3$, $X_2 \cdot X_4$, and $X_3 \cdot X_4$ in the model and is useful in food product design.

In a similar way, different kinds of 2^{n-p} fractional factorial designs can be constructed. In extreme cases, all the columns for estimating interactions are "sacrificed" and used to assign new factors. A 2^{n-p} experimental design would be able to check the significance of $(2^{n-p} - 1)$ variables. Such 2^{n-p} designs are characterized as a saturated fractional factorial designs and are usually used in the screening stage (Section 3.2.2.2). A saturated design does not estimate interaction between factors and supplies an alternative approach to the

TABLE 3.7. A 2^{5-1} Fractional Factorial Design.

No.	2^4 Factorial Design				$= X_1 \cdot X_2 \cdot X_3$
	X_1	X_2	X_3	X_4	X_5
1	-1	-1	-1	-1	-1
2	-1	-1	-1	$+1$	-1
3	-1	-1	$+1$	-1	$+1$
4	-1	-1	$+1$	$+1$	$+1$
5	-1	$+1$	-1	-1	$+1$
6	-1	$+1$	-1	$+1$	$+1$
7	-1	$+1$	$+1$	-1	-1
8	-1	$+1$	$+1$	$+1$	-1
9	$+1$	-1	-1	-1	$+1$
10	$+1$	-1	-1	$+1$	$+1$
11	$+1$	-1	$+1$	-1	-1
12	$+1$	-1	$+1$	$+1$	-1
13	$+1$	$+1$	-1	-1	-1
14	$+1$	$+1$	-1	$+1$	-1
15	$+1$	$+1$	$+1$	-1	$+1$
16	$+1$	$+1$	$+1$	$+1$	$+1$

Plackett-Burman design to screen variables at the initial stage of food product design. The computer program STATISTICA has a good functionality to generate different kinds of 2^{n-p} designs with specific model forms.

3.4.4 EXPERIMENTAL DESIGNS FOR MODELS OF THE SECOND DEGREE

2^n factorial designs, or its variants, only support building of a linear model with or without interactions between the factors. Actually, there are systems in which a significant nonlinear relationship exists between the influencing variables and the food quality indices in the researched range. Such a relationship cannot be estimated and described by a linear model with adequate precision. In this case, models of second degree supported by specific experimental designs should be used. The Taylor expansion equation of the second degree [Equation (3.7)] is usually suitable to model such food systems. This quadratic model takes into consideration the effects of each factor, the interactions between and among the factors, and the curvature. There are several kinds of statistical experimental designs supporting the establishment of a quadratic model, namely, the estimation of the effects of each factor, the interactions between and among the factors, and the curvature. They are introduced in the following sections of this chapter.

3.4.4.1 3^n Factorial Design

Regarding the trial number, a 3^n factorial design is suitable for supporting the building of a quadratic model, if there are less than four significant variables ($n \leq 4$) selected for modeling in the food system. Similar to a 2^n factorial design, in a 3^n factorial plan n is the independent variable number, 3 the three test levels (high, middle, and low levels) of each variable, and 3^n equals the total number of trials. A 3^n experimental design supplies 3^n degrees of freedom, in which one is fixed for determining the total average value β_0 (constant term) in the model. The remaining $(3^n - 1)$ degrees of freedom then allow estimation and calculation of the effects of each factor, the interactions between and among factors, and the curvature in the system. However, the number of trials increases drastically with the number of variables. For four variables, a plan of $3^4 = 81$ trials would be needed! In practice to perform so many trials is sometimes impossible and usually uneconomical.

A 3^n factorial design is constructed by the combination of all the possible test levels of each variable. It can be divided into four subgroups:

(1) a 2^n factorial plan with 2^n trials
(2) $3 \cdot n$ central points of all the surfaces
(3) border middle points
(4) one central point (in practice, this should be repeatedly performed)

TABLE 3.8. Coded 3^3 Factorial Design.

No.	X_1	X_2	X_3	No.	X_1	X_2	X_3	No.	X_1	X_2	X_3
1	−1	−1	−1	9	−1	+1	0	21	−1	0	0
2	−1	−1	+1	10	−1	−1	0	22	+1	0	0
3	−1	+1	−1	11	+1	+1	0	23	0	−1	0
4	−1	+1	+1	12	+1	−1	0	24	0	+1	0
5	+1	−1	−1	13	−1	0	+1	25	0	0	−1
6	+1	−1	+1	14	−1	0	−1	26	0	0	+1
7	+1	+1	−1	15	+1	0	+1	Surface central			
8	+1	+1	+1	16	+1	0	−1	symbol ★			
2^3 design				17	0	−1	+1	27	0	0	0
symbol ●				18	0	−1	−1	Overall central			
				19	0	+1	−1	symbol ✿			
				20	0	+1	−1				
				Border middle points							
				symbol ▲							

In Table 3.8 is a coded 3^3 factorial plan which is graphically presented in Figure 3.6. In Table 3.8 "−1", "0", and "+1" correspond to the real low, middle, and high levels of each factor. The middle level of one factor is the average value of the sum of the low and high levels of this factor. All trials in the plan are classified into the above four categories and are illustrated with different symbols. It can be concluded from the graphic that the trials are distributed more densely in a 3^n factorial design than in a 2^n plan. These more densely distributed supporting points would ensure the determination of the quadratic effects of each variable and the building of a quadratic model. Other 3^n factorial designs for two to four variables are listed in the appendix.

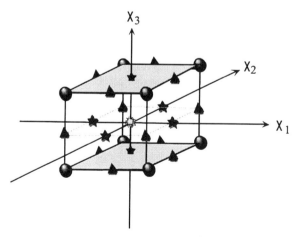

Figure 3.6 Graphical description of the 3^3 factorial design.

Similar to 2^{n-p} designs, there are fractional factorial 3^{n-p} designs. Two types of fractional factorial 3^{n-p} are well known. One is the standard Box-Behnken designs published by Box and Behnken in the 1960s. They were constructed by combining two-level factorial designs with incomplete block designs and have complex confounding of interaction. The other types are the standard designs enumerated by Connor and Zelen for the National Bureau of Standards of the U.S. Department of Commerce in 1980s. Box-Behnken designs are economical and particularly useful when it is expensive to perform many experimental runs. STATISTICA and STATGRAPHICS have the functionality of generating both types of fractional factorial 3^{n-p} designs.

3.4.4.2 Mixed $2^n 3^m$ Factorial Designs

As discussed above, a 3^n factorial design can support estimation of the quadratic effects of all the independent variables. In practical food product design, there are systems in which some of the variables have only linear effects on the target parameters, whereas others have quadratic influences. In this case, the full 3^n factorial design can be simplified into a mixed $2^n 3^m$ factorial design, with a reduced trial number to determine the relationship between all the independent variables and the food quality objectives. The structure of a mixed $2^n 3^m$ factorial design is actually constructed by mixing a 2^n with a 3^m factorial designs, where n means the number of variables to be tested in a linear way, and m the number of variables with quadratic effects.

Connor and Young enumerated a serial of full and fractional factorial $2^n 3^m$ designs for the National Bureau of Standards of the U.S. Department of Commerce in the 1980s. In a full factorial $2^n 3^m$ design, the main, interaction, and quadratic effects of any factor are orthogonal and can be estimated. However, the main effects for these mixed level fractional factorial designs are not necessarily orthogonal.

In practice, factors whose nonlinear effect is not significant are arranged in the 2^n part, whereas those with distinct quadratic effects will belong to the 3^n part. This kind of design can also be used, if category variables such as type of flour or type of machinery are included in the study, which can only be tested in two or three levels. An example of a mixed $2^2 3^1$ factorial design for three factors X_1, X_2, and X_3 is given in Table 3.9 and shown in Figure 3.7. The linear main and interactive effects of X_1 and X_2 can be tested, whereas the quadratic effect of X_3 can be additionally estimated through its three levels. The intended model form is given in Equation (3.12). It is similar to a linear model with interaction terms, except for one additional term of $\beta_{33} \cdot X_3^2$, which describes the quadratic effect of X_3 in the system.

$$Y = \beta_0 + \beta_1 \cdot X_1 + \beta_2 \cdot X_2 + \beta_3 \cdot X_3 + \beta_{12} \cdot X_1 \cdot X_2$$
$$+ \beta_{13} \cdot X_1 \cdot X_3 + \beta_{23} \cdot X_2 \cdot X_3 + \beta_{33} \cdot X_3^2 \qquad (3.12)$$

TABLE 3.9. A Coded Mixed $2^2 3^1$ Factorial Design.

No.	X_1	X_2	X_3	Symbol
1	-1	-1	-1	
2	$+1$	-1	-1	
3	-1	$+1$	-1	
4	$+1$	$+1$	-1	●
5	-1	-1	$+1$	
6	$+1$	-1	$+1$	
7	-1	$+1$	$+1$	
8	$+1$	$+1$	$+1$	
9	-1	-1	0	
10	$+1$	-1	0	☼
11	-1	$+1$	0	
12	$+1$	$+1$	0	

3.4.4.3 Central Composite Designs (CCD)

3.4.4.3.1 Introduction

The CCD is the foundation of the well-known response surface methodology (RSM) and is used to estimate parameters of a full second-degree model in all scientific research areas. One of its main advantages is that a CCD can be constructed in a sequential program of experimentation by building onto information gathered previously from a 2^n factorial design. In other words, if

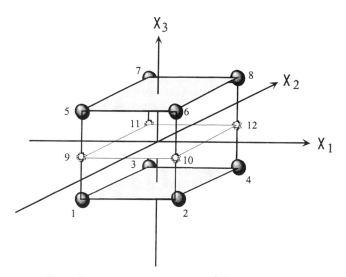

Figure 3.7 Graphical description of a $2^2 3^1$ factorial design.

a linear model based on a 2^n factorial design turns out to be insignificant, then some extra trials can be designed, according to the principles of a CCD, to repair the model. All these data will be used to build a quadratic model. This is also known as the buildup principle of the CCD. Normally, a quadratic model would meet the needs for accuracy in practical product development.

3.4.4.3.2 Design Construction

In a CCD, each independent variable is tested on five levels. From this point of view, a CCD is a strongly reduced full 5^n factorial experimental design. A CCD is constructed from three parts. The first is a cube portion of one 2^n factorial design or a 2^{n-p} fractional factorial design. The second is $2 \cdot n$ combinations of two test levels at the so-called star points or axial points (usually characterized as codes $\pm \alpha$) of all the independent variables that allow the estimate of quadratic coefficients in the model. The third parts are replications at the overall central point.

(1) Cube points n_c: 2^n factorial design or 2^{n-p} fractional factorial design (cube corner points trials)
(2) Star points n_s: $2 \cdot n$ star points trials
(3) Center points r: trials at the overall central point (all the factors are set at their middle level)

Thus, the trial number of a CCD is usually the sum of cube, star, and center points. For simplicity, the number of factors is usually limited to five or six.

The first part of a CCD plan is normally a full 2^n factorial design as stated in Section 3.4.3.1. The trials on the star points should be arranged in such a way that in each trial only one variable is set at its coded level "$-\alpha$" or "$+\alpha$", whereas the others are at their coded middle value "0." To examine the experimental error, the trial at the overall central point usually needs to be repeated three to six or even more times. The example in Table 3.10 is a CCD for two factors X_1 and X_2 ($n = 2, r = 6$) which is illustrated in Figure 3.8 as well.

3.4.4.3.3 Calculation of α Value

As stated in Section 3.4.1, the orthogonality, which is the prerequisite of a statistical experimental design regarding the independent information, can be extracted from the design. Technically, two columns in a design matrix are orthogonal if the sum of the products of their elements within each row is equal to zero. The orthogonality of a CCD is attained by arranging a definite number of trials repeated at the overall central points and/or by choosing a suitable axial distance α (alpha) value for level calculation at the star points. α is the distance of the star points from the center of the design and is a coded value.

TABLE 3.10. CCD for Two Factors with Six Repeated Trials at Central Point.

	No.	X_1	X_2
2^2 full factorial design ●	1	−1	−1
	2	−1	+1
	3	+1	−1
	4	+1	+1
Trials at star points ★	5	−α	0
	6	+α	0
	7	0	−α
	8	0	+α
Repeated trials at central point ✷	9	0	0
	10	0	0
	11	0	0
	12	0	0
	13	0	0
	14	0	0

−α and +α are the coded levels of each variable at its star points. The α value is calculated according to the formula given below [Equation (3.13)], so that the orthogonality of the whole CCD experimental design would be retained.

$$\alpha = \frac{\left[\left[\sqrt{n_c + n_s + r} - n_c^{1/2}\right]^2 \times n_c\right]^{0.25}}{\sqrt{2}} \tag{3.13}$$

In most cases the full 2^n factorial design is used in the cube portion in a CCD and thus the α value is calculated with the following formula:

$$\alpha = \frac{\left[\left[\sqrt{2^n + 2 \times n + r} - 2^{n/2}\right]^2 \times 2^n\right]^{0.25}}{\sqrt{2}}$$

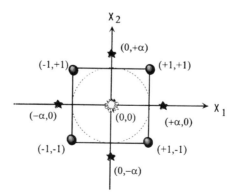

Figure 3.8 Graphical description of a CCD for two variables.

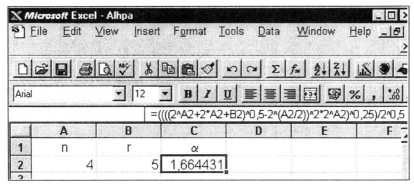

Figure 3.9 Calculation of the α value with Ms-Excel$^®$.

The formula itself is somewhat complicated. However, with the help of a small computer program such as Ms-Excel$^®$ the α value can easily be calculated. Using Ms-Excel$^®$ the values of n and r can be input into cells A2 and B2, respectively, and cell C3 will show the calculated α value (Figure 3.9). The equal mark "=" is typed first in cell C3 and then the Equation (3.13) in a form that is accepted by the computer program. It is given here as follows:

$$= ((((2\char`^A2 + 2*A2 + B2)\char`^0.5 - 2\char`^(A2/2))\char`^2*2\char`^A2)\char`^0.25)/2\char`^0.5$$

Pressing the RETURN key will let the α value appear in cell C3. For example, if n and r were 4 and 5, respectively, then the α value would be 1.664 (Figure 3.9). This Ms-Excel$^®$ program is saved under the file name "Alpha.xls". For any further calculations of α values, one only needs to replace the n and r values in cell A2 and B2; the α value would then be calculated and shown automatically.

Rotatability is another important characteristic of a CCD. Rotatable designs focus on how to extract the maximum amount of (unbiased) information about the dependent variable from the experimental region of interest and leave the least amount of uncertainty for the prediction of future values. Any kind of rotation of the original design points will generate the same amount of information, that is, generate the same information function. By adding star points or axial points to the simple 2^n factorial design points, one can achieve rotatable and often orthogonal or nearly orthogonal designs. The rotatable characteristic depends on the number of center points in the design and on the α (alpha). It can be shown that a design is rotatable if α is $n_c^{1/4}$. To make a CCD both (approximately) orthogonal and rotatable, one would first choose the axial distance for rotatability and then add center points so that the orthogonality is achieved.

3.4.4.3.4 Transformation between Coded and Real Values

The real values of test levels at the star points for each independent variable can be calculated according to the following transformation formulas [Equations (3.14) and (3.15)]:

$$X_{real.-\alpha} = \frac{X_{real.upper} + X_{real\,low}}{2} - \frac{X_{real.upper} - X_{real.low}}{2} \times \alpha \qquad (3.14)$$

$$X_{real.+\alpha} = \frac{X_{real.upper} + X_{real.low}}{2} + \frac{X_{real.upper} - X_{real.low}}{2} \times \alpha \qquad (3.15)$$

where

$X_{real.-\alpha}$: real value of lower level at star point (corresponding to $-\alpha$)
$X_{real.+\alpha}$: real value of upper level at star point (corresponding to $+\alpha$)
$X_{real.low}$: real lower level corresponding to coded levels -1
$X_{real.upper}$: real upper level corresponding to coded levels $+1$

To get the real experimental design for trial, one only needs to rewrite the coded design, in which the $-\alpha$, -1, $+1$, and $+\alpha$ correspond to the real levels $X_{real.-\alpha}$, $X_{real.upper}$, $X_{real.lower}$, and $X_{real.+\alpha}$. Such a transformation can also be atomized in table calculation programs such as Ms-Excel® and Lotus 1-2-3®. The procedure is similar to that shown in Figure 3.9.

3.4.4.3.5 A Practical Example

A conventional dry heat treatment process was used to inactivate the trypsin inhibitor in soybean flour. The temperature T (°C) and the time t (min) of heat treatment are selected for investigation. The research range of interest for T is from 70 to 100°C, and for t from 20 to 60 min. In this case, these four boundary values should be adopted as $-\alpha$ and $+\alpha$ levels so that the research range is located within the intended range, because the α value is higher than the absolute coded upper or lower level 1.

In this system, the α value is 1.32 and is calculated according to Equation (3.13) with r of 6. To calculate the real level for T and t at their lower and upper levels, Equation (3.8) should be modified as given below [Equation (3.16)]. Now 70°C for T and 20 min for t correspond to the coded level -1.32, and 100°C for T and 60 min for t correspond to the coded level $+1.32$.

$$X_{i.real} = \frac{X_{real.+\alpha} + X_{real.-\alpha}}{2} + \frac{X_{real.+\alpha} - X_{real.-\alpha}}{X_{code.+\alpha} - X_{code.-\alpha}} \times X_{i.code} \qquad (3.16)$$

TABLE 3.11. CCD for Two Variables T and t.

No.	T	t
1	74	25
2	74	55
3	96	25
4	96	55
5	70	40
6	100	40
7	85	20
8	85	60
9–14	85	40

where $X_{i.\text{real}}$ is the real value of any coded value $X_{i.\text{code}}$ (here $+1$ and -1). The real values of T and t at lower (-1) and upper $(+1)$ levels are:

$$\text{For } T: \quad T_{-1.\text{real}} = \frac{100 + 70}{2} + \frac{100 - 70}{1.32 - (-1.32)} \times (-1) \approx 74°C$$

$$T_{+1.\text{real}} = \frac{100 + 70}{2} + \frac{100 - 70}{1.32 - (-1.32)} \times (+1) \approx 96°C$$

$$\text{For } t: \quad t_{-1.\text{real}} = \frac{60 + 20}{2} + \frac{60 - 20}{1.32 - (-1.32)} \times (-1) \approx 25 \min$$

$$t_{+1.\text{real}} = \frac{60 + 20}{2} + \frac{60 - 20}{1.32 - (-1.32)} \times (+1) \approx 55 \min$$

The experimental design for trial would then be transformed from Table 3.10 and is listed in Table 3.11.

3.4.5 D- AND A-OPTIMAL DESIGNS

3.4.5.1 Introduction

The factorial designs discussed in this chapter are not always applicable for some food processes because of their functional restrictions. A 2^n factorial experimental design is constructed on the basis of the assumption that, in the studied space, the relationship between the response and variables is linear. Every trial in a factorial design must be performed, and the trial number increases rapidly beyond the economical affordable limit with a large number of variables. Although in the CCD only a small number of trials should be performed, however, it requires the exact setting of the test levels at that defined values and cannot be changed. In addition, the trials in all factorial experimental designs locate extremely at the boundaries of the researched room.

To overcome these shortcomings, the D-optimal design is developed. In a D-optimal design, the test level of each variable can be selected flexibly, namely, a variable can be tested at as many levels as the food researcher wants. The number of test levels of different variables can be different or the same. The expected orthogonality of the whole experimental design can be possibly achieved through selection of designs. An essential consideration of using the D-optimal design is that the total trial number in the final D-optimal design must be bigger than the number of coefficients in the intended regression model. Usually, the trial number in the design should be equal to or more than 1.5 times of the coefficients number.

In general, D-optimal designs are computer-generated designs following different computation algorithms. "D-optimal" means that these designs are selected from a list of valid candidate runs that provide as much orthogonality between the columns of the design matrix as possible. They can accordingly extract the maximum amount of information from the experimental region with respect to a stated model and the run number for the experiment. The maximum amount of information regarding a dependent variable from the experimental region can be extracted, if all factor effects in the design are orthogonal to each other.

3.4.5.2 D-Optimality and A-Optimality

The determinant D of a square matrix is a specific numerical value that reflects the amount of independence or redundancy between the columns and rows of the matrix. This basic relationship also extends to larger design matrices. That is, a large determinant value represents independence and a small determinant value (approaching zero) represents redundancy. Finding a design with maximally independent factor effects is accomplished by maximizing the determinant of the design matrix. This criterion for selecting a design is called the D-optimality criterion. Actually, the computations are commonly performed on a simple cross-product matrix, instead of on correlation matrix of vectors. In matrix notation, if the design matrix is denoted by X, then the quantity of interest is the determinant of $X'X$ (X-transposed times X). Thus, the search for D-optimal designs aims to maximize $|X'X|$, where the vertical lines ($|..|$) indicate the determinant. Another way to look at the issue of independence is to maximize the diagonal elements of the $X'X$ matrix while minimizing the off-diagonal elements. The so-called trace criterion or A-optimality criterion expresses this idea. D-optimal designs minimize the expected prediction error for the dependent variable. In other words, D-optimal designs maximize the precision of prediction, and thus, the information (which is defined as the inverse of the error) that is extracted from the experimental region of interest.

3.4.5.3 Constructing Optimal Designs

D- or A-optimal designs are constructed by searching for n runs from a candidate list that specifies which regions of the design are valid or feasible, given a user-specified number of runs for the final experiment. This candidate trial set is the discrete set of "all potential good runs" and is generated through combining all the levels of each variable. It consists of one or more of the following sets of points according to the model form or terms in the model:

- the extreme vertices of the constrained region
- the centers of the edges
- the centers of the various high dimensional faces
- the overall centroid

This "searching for" the best design is not an exact method, but rather, an algorithmic procedure that uses certain search strategies to find the best design according to the respective optimality criterion. The most popular searching algorithms used in experimental design computer programs are as follows:

(1) Sequential or Dykstra method: It starts with an empty design, searches through the candidate list, and chooses the one that maximizes the specified criterion in each step. There are no iterations involved; the programs simply pick the requested number of points sequentially. Thus, this method is fast, but often most likely to fail by yielding a design that is only locally optimal.

(2) Simple exchange (Wynn-Mitchell) method: This method starts with an initial design of the requested size (constructed via the sequential search algorithm described above). In each iteration, one run in the design is dropped, and another is added from the list of candidate points. At each step, the point that contributes least with respect to the chosen optimality criterion (D or A) is dropped from the design; then the algorithm adds a point from the candidate list to optimize the optimality. The algorithm stops when no further improvement is achieved with additional exchanges.

(3) DETMAX algorithm (exchange with excursions): This algorithm is probably the best known and most widely used optimal design search algorithm. Like the Simple exchange method, an initial design is constructed via the sequential search algorithm, and then the search begins with a simple exchange as described above. However, if the respective criterion (D or A) does not improve, the algorithm will undertake excursions. Specifically, the algorithm will add or subtract more than one point at a time, so that, during the search, the number of points in the design may differ from the requested design size. The iterations will stop when the chosen criterion (D or A) no longer improves within the maximum excursion defined by the user.

(4) Fedorov (simultaneous switching): This is the simultaneous switching method proposed by Fedorov. In each iteration, all possible pairs of points in the design and those in the candidate list are evaluated. The algorithm will then exchange one pair to optimize the design for the chosen criterion. Thus, it is easy to see that this algorithm can be somewhat slow because of the large amount of comparison operations required to exchange a single point.

(5) Modified Fedorov (simultaneous switching): This algorithm represents a modification of the basic Fedorov algorithm. It also begins with an initial design of the requested size (constructed via the sequential search algorithm). In each iteration, the algorithm will exchange each point in the design with one chosen from the candidate list to optimize the design according to the chosen criterion (D or A). Unlike the simple exchange algorithm described above, the exchange is not sequential, but simultaneous. Thus, in each iteration, each point in the design is compared with each point in the candidate list, and the exchange is made for the pair that optimizes the design. The algorithm terminates when there are no further improvements in the respective optimality criterion. If time is not a consideration, the modified Fedorov algorithm is the best method to use.

In summary, there are usually no exact solutions to the optimal design problem. It is possible to have several "local minima" for the chosen optimality criterion. For example, at any point during the search a design may appear optimal unless half of the runs in the design are simultaneously discarded and certain other points from the candidate list are chosen. If only individual points or a few points are exchanged, then no improvement occurs. Therefore, it is important to try a number of different initial designs and algorithms. If the same or similar optimal design results are obtained after applying several of the search algorithms, one can be reasonably sure that the design is an overall minimum or maximum, not just a local minimum or maximum.

3.4.5.4 Design Efficiency

D-efficiency, A-efficiency and G-efficiency are standard measures that summarize the efficiency of a D-optimal design. D-efficiency can be interpreted as the relative number of runs (in percent) that would be required by an orthogonal design to achieve the same value of the determinant $|X'X|$. However, it is only a theoretical "yardstick," because an orthogonal design may not be possible in many cases. Therefore, D-efficiency should be used as a relative indicator of efficiency, to compare other designs of the same size that are constructed from the same candidate list. In addition, this measure is only meaningful if the factor settings in the candidate list and the final design are recoded at a minimum of -1 and a maximum of $+1$.

A-efficiency can be interpreted as the relative number of runs (in percent) that would be required by an orthogonal design to achieve the same value of the trace of $(X'X)^{-1}$. Again, this measure is a relative indicator of efficiency and is only meaningful if the factor settings in the design are recoded to the -1 to $+1$ range.

G-efficiency is related to the so-called G-optimality criterion. G-optimal designs are defined as those that will minimize the maximum value of the standard error of the predicted response.

3.4.5.5 Special Applications

The D-optimal design features can be used to "repair" or enhance an existing design. For example, suppose we ran an orthogonal design, but some of the data were lost, some effects of interest could no longer be estimated, and it was not feasible to repeat any of the lost runs. In this case, we can set up the candidate list from among all valid points for the experimental region, add all the runs that we have completed to the list, and instruct the program to force those points into the final design (forced inclusion). The program will then only exclude the points we did not actually run. In this manner, we can find the best single run to add to an existing experiment to optimize the respective criterion. A set of experimental runs of interest can also be simply specified as "Inclusions." It means that these runs of interest are reserved in the final optimal design by defining them as Inclusions. In this way the researcher has the flexibility to make a design for model improvements, such as to deconfound interactions or to estimate curvature. Most experimental design computer programs such as STATISTICA and MODDE can handle such problems.

D-optimal designs are particularly useful when the research region is subject to complex constraints and one wants to fit particular models with the least number of runs. For such kinds of experimental problems, there are usually no classical designs to apply. For example, linear constraints on the factor settings will reduce the experimental region to an irregular polyhedron. In this case, D-optimal design is the preferred choice because it makes efficient use of the entire experimental space.

In case the number of experimental runs affordable is smaller than the number of runs of any available classical design, D-optimal design can also be applied to reduce the number of runs when it is not feasible to run the number of trials required for a classical experimental design.

3.5 REALIZATION OF EXPERIMENTAL DESIGNS

The next step following the experimental planning is to perform the design. Before conducting trials, the experiments in the design should be randomized, so that the correlation of experimental results with time and other unknown influencing factors is minimized. Through randomization, similar trials are placed in such a sequence that they are set apart from each other as far as possible. All the experimental software provides such randomization treatment of a generated design. However, this action may not be feasible for some practical experimentation. Sometimes blocking or grouping experiments that have specific requirements, such as a specific piece of equipment, helps reduce the experimental error.

While conducting the experiments, the levels of each variable must be kept as close as possible to the value designated in the design. In practice, there are always deviations between actual and design values. Therefore, it is extremely important that the actual or true level value of each variable is documented honestly and in detail. Only these actual experimental level values can be used to build a model. Thus, any strong deviation of actual levels from the desired setting levels may weaken the expected orthogonality of the design and cause inexact optimizations.

Based on the same experimental design, an unlimited number of food quality objectives (responses) could be tested and modeled. Each quality index of the food product from each trial is determined and analyzed, and a corresponding coded or real model is built.

3.6 MODEL BUILDING

Building a model is one of the most important steps in food product development. After the experiments have been performed and the data collected, the intended model is fitted to the data by using regression analysis. Building a model actually consists of establishing the quantitative relationships between the independent variables and the target parameter according to the experimental data. In other words, the purpose of this step is to obtain estimates for each coefficient β_0, β_j, β_{jk}, and/or β_{jj} in the model that was proposed before starting experimentation. This is the most critical step to achieve success in food product design. Before personal computers and suitable computer programs were readily available, this step was intimidating and seen as the largest obstacle for using statistical experimental product design. Nowadays, the situation has completely changed, and the regression can be performed with great ease if proper computer software is at hand.

3.6.1 BASIC THEORY OF REGRESSION

3.6.1.1 Principle of Least Square

Regression analysis is an important statistical and mathematical method. Its principles are simple and straightforward and can be found in any book on statistical mathematics. There is, therefore, no need to explain them here.

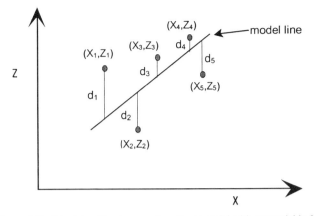

Figure 3.10 Principle of least square for a linear model with one variable X.

As already discussed in Section 3.3 of this chapter, a single set of data can be described with several different mathematical models but with different fit quality. Among all these possible models, only one or two fit the data set most closely. Thus, the selection of an adequate model for regression is extremely important. The task of regression is to estimate the model coefficients to provide the most precise description of the original actual experimental data. To do this, the least square principle is used. Figures 3.10 and 3.11 demonstrate this principle for a linear model with one and two independent variables, respectively. In Figure 3.10 the thick line represents the model fitted based on five sets of data (X_1, Z_1), (X_2, Z_2), (X_3, Z_3), (X_4, Z_4), and (X_5, Z_5). The distances of the responding Z from each of the original data points to the model

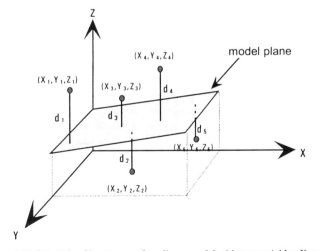

Figure 3.11 Principle of least square for a linear model with two variables X and Y.

line are d_1, d_2, d_3, d_4, and d_5, respectively. The best model to fit these five sets of data must fulfill the condition of "$d_1^2 + d_2^2 + d_3^2 + d_4^2 + d_5^2 = $ minimum." In this way the regression can estimate or find the coefficients of the best model. Figure 3.11 demonstrates the same principle in a system of two variables X and Y. In this case d_i is the distance from every original point (X_i, Y_i, Z_i) to the plane of the model.

3.6.1.2 Estimation of Linear and Interaction and Quadratic Terms

Regression analysis ensures that the model obtained will be the most accurate equation for describing the data. This equation has practical meaning, and its parameter values are important. To do regression, there must be a set of precise original experimental data, a definite equation, and a calculation criterion for determining the coefficients. If the experimental data contain a large amount of error, the model will not accurately describe the relationship between the food quality and influencing factors. In other words, the accuracy of the experimental data cannot be improved through regression.

In regression, the actual experimental data (not coded) are usually used to build a real model. In such a model, all independent and dependent variables are actual values, and the coefficient sizes are related to the dimension of each variable. Accordingly, the importance of each term cannot be judged directly from its coefficients. However, the relative importance of each term can be determined according to the standardized regression coefficients (β), which are usually supplied by standard computer programs such as SPSS/PC+, SAS, STATISTICA, MODDE, and SigmaStat. The larger the standardized coefficient, the larger the effect of the corresponding term on the dependent variable. Occasionally however, the coefficient size does not directly correspond to the importance of the variable. This point will be explained in detail in Section 3.7.

Coded experimental data can be used for regression modeling, if actual experimental data are coded to the range of -1 to $+1$ according to Equation (3.9). In other words, coded experimental levels and actual response data are used to build a coded model. In a coded model, all variables are standardized and, therefore, dimensionless. Accordingly, a term with a larger coefficient is of greater importance than the one with a smaller coefficient.

In regression, all terms in the model are treated as different independent variables. For example, the interaction effect $X_j \cdot X_k$ is considered as a new independent variable, and its coefficient B_{ij} is estimated on the basis of a new column of data (levels) in the design obtained by multiplying the actual test levels of variable X_j by X_k ($X_j \cdot X_k$). Similarly, a quadratic term $B_{jj} \cdot X_j^2$ is considered the effect of a new variable X_j^2 as well. However, its coefficient B_{jj} can only be correctly determined by using a column in which each level is built according to the following formula [Equation (3.17)]:

$$\left(X_j^2\right)_i = [(X_j)_i]^2 - \frac{\sum_{i=1}^{m}[(X_j)_i]^2}{m} \tag{3.17}$$

where m is the total trial number and i the row number. In this way, the linear effect of X_j will be separated from that of the quadratic term X_j^2. Such conversion can be quickly performed with help of a table calculation program such as MS-Excel®. Most of the modern statistical software can calculate all the coefficient estimates automatically.

3.6.2 EXPERIMENTAL DESIGNS AND SUPPORTED REGRESSION MODELS

Before starting regression, the model form must be selected. In fact, the model form is defined in the stage of generating the experimental design. Different statistical experimental designs support the building of models of different degrees. Usually, a set of experimental data supporting the assessment of a quadratic model would also support a model of a lower degree. Table 3.12 summarizes the statistical experimental designs and the corresponding models supported by them.

TABLE 3.12. Various Experimental Designs and Their Supported Models.

Experimental Design	Model Form
2^{n-p} Fractional factorial design; Plackett-Burman design	$Y = B_0 + \sum_{j=1}^{n} B_j \cdot X_j$
2^n Factorial design	$Y = B_0 + \sum_{j=1}^{n} B_j \cdot X_j$ $Y = B_0 + \sum_{j=1}^{n} B_j \cdot X_j + \sum_{j<k=2}^{n} B_{jk} \cdot X_j \cdot X_k$
3^n factorial design Central composite design D-optimal design	$Y = B_0 + \sum_{j=1}^{n} B_j \cdot X_j$ $Y = B_0 + \sum_{j=1}^{n} B_j \cdot X_j + \sum_{j<k=2}^{n} B_{jk} \cdot X_j \cdot X_k$ $Y = B_0 + \sum_{j=1}^{n} B_j \cdot X_j + \sum_{j<k=2}^{n} B_{jk} \cdot X_j \cdot X_k + \sum_{j=1}^{n} B_{jj} \cdot X_j^2$

3.6.3 REGRESSION WITH SOFTWARE

Regression is a complex mathematical calculation. This is one of the main reasons why the techniques of mathematical modeling have been quite limited in the past. Because of the development of personal computers and statistical software packages, regression calculations can now be performed quickly and easily with high precision. As a user of these standard programs, one does not need to know how the coefficients are calculated. All one needs to know is how to enter the experimental data into a computer and start the regression procedure. The typical output data obtained from regression analysis includes:

- coefficients estimates and related statistics
- analysis of variance (ANOVA)
- diagnostic tests to assess the adequacy of the model (F-value , R^2, Adj. R^2, and Q)
- list of actual and predicted values for each trial (residual statistics)
- correlation coefficients matrix between all the variables in the model

Some standard computer programs such as SPSS/PC+, SAS, and STATISTICA still supply information about the standardized regression coefficients, which can be helpful in understanding the relative importance of the corresponding model variables. Special attention should be paid to those variables that contribute most to the response.

Multiple regression analysis is usually used by standard software to establish the most suitable coefficients in the model based on the actual experimental data. Essentially, there are two effective multiple regression procedures that are commonly used. One procedure, known as forward multiple stepwise regression, begins without any variables in the model. Variables are added to the regression equation in order of their contribution to the prediction of the dependent variables providing that they contribute at a specified level of significance. The significance level, which can be chosen by the investigator, is normally set between $p = 0.15$ and $p = 0.10$ for adding variables and at $p \leq 0.10$ or $p \leq 0.15$ for removing variables from the model. The other procedure is the backward multiple stepwise regression. In this procedure, all the variables are initially entered into the equation (full model); then, the one that contributes the least to the prediction is removed. The variable removal process is continued until removing a succeeding variable significantly reduces the reliability in predicting the dependent variable.

SAS, SPSS, STATISTICA, BMDP, and MiniTable are the most widely used statistical computer programs or packages. When using these programs to do regression, it is sometimes necessary to write a macro (a series of commands necessary for data analysis). Writing a macro can be complicated. Table 3.13

TABLE 3.12. Regression Macro for SPSS/PC+.

Macro	Explanation of the Steps
Data list file='a:\ hu_test1.dat'free/	Input of experimental data from the file name hu_test1.dat (ASCII Format) saved on disk A:.
x1 x2 x3 x4 y.	Definition of variables.
Compute x1x2 = x1*x2.	Generate new variables x1x2, x1x3, x1x4, x2x3,
Compute x1x3 = x1*x3.	x2x4, and x3x4 according to these formulas.
Compute x1x4 = x1*x4.	(Quadratic terms can also be generated in this way)
Compute x2x3 = x2*x3.	
Compute x2x4 = x2*x4.	
Compute x3x4 = x3*x4.	
Set listing='a:\ hu_test1.lis'.	Output the regression result to the file hu_test.lis which would be saved on disk A:.
Regression variables=x1 to x3x4 x3x4/dependent=y/ method=backward/ casewise=all.	Define x1, x2, x3, x4, x1x2, x1x3, x1x4, x2x3, x2x4, and x3x4 as independent and *y* as dependent variables; select the backward stepwise regression method (an alternative forward or all can be used). The last switch casewise=all describes the extent of regression analysis.
Correlation variables=x1 to y.	Calculates a Pearson correlation between all the variables.
finish.	Finish the macro and go back to DOS.

shows a macro that has been used successfully to do regression in the SPSS/PC+ environment.

Along with the advance of modern computer operating systems with a graphic interface such as Microsoft Windows®, many modern statistical computer programs have been developed. They usually run on a personal computer with an Intel 80386 processor or higher, and the regression program is convenient and straightforward. Usually, all one needs to do is to type, copy, or import the experimental data into a spreadsheet, define the variables, and select the regression command from the menu or toolbar. The regression will be calculated, and the results sent to the screen and/or a file. Some popular software packages and corresponding macros for doing regression are listed below, and the most popular software houses are listed at the end of Chapter 1.

3.6.3.1 Statistical Analysis System

Statistical Analysis System (SAS) is one of the most widely used computer statistical packages. Although originally it was only available on mainframe

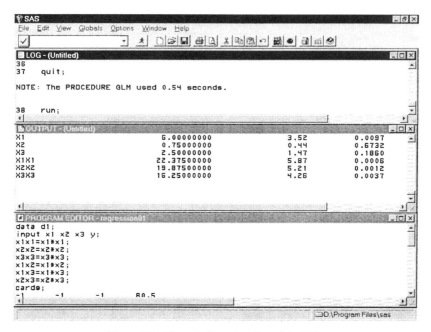

Figure 3.12 The typical user interface of the SAS.

computers, personal computer versions are now available. Figure 3.12 is the typical user interface of SAS for Windows. In the PROGRAM EDITOR subwindow the macro is edited, the LOG window keeps a record, and the OUTPUT window shows the results of macro execution.

The procedures PROC GLM or PROC RSREG use the least squares method to fit general linear models. The PROC GLM is suitable for solving general regression tasks, whereas PROC RSREG is specialized for response surface problem. The former procedure has the advantage of allowing different modifications of the model form, especially the elimination of the insignificant terms from the model. On the contrary, the PROC RSREG is based on a full quadratic model and provides more information related to RSM. Statistical methods available in these procedures are regression, ANOVA, analysis of covariance, multivariate analysis of variance, and partial correlation. These procedures can also be run interactively. Table 3.14 shows an example of a macro for performing regression using the procedure PROC GLM for a response Y with double trials Y1 and Y2. Obviously, the running results (Table 3.15) of this macro show that these terms X2, X1*X2, X2*X3, and X2*X4 (underlined in the table) are not significant in the model and therefore should be eliminated from the model. These variables are eliminated by rewriting the model statement as "Model y = x1 x3 x4 x1*x3 x1*x4 x3*x4." The renewed running of the improved

TABLE 3.13. Regression Macro for SAS.

Macro	Explanation
data;	Data definition
input dpoint x1 x2 x3 x4 y1 y2;	Define independent variables x1, x2,
array ys{2} y1–y2;	x3, x4 and response y with double
do i = 1 to 2;	trials y1, y2 as well as their
i = ys{i}; 7	corresponding experimental data
output;	
end;	Finish the data definition
keep x1 x2 x3 x4 y;	
cards;	Input experimental data for x1–x4
1 1 1 1 1 3.62 3.67	and y1–y2
2 1 1 1 -1 2.54 2.65	
3 1 1 -1 1 7.99 7.88	
4 1 1 -1 -1 3.69 3.55	
5 1 -1 1 1 2.82 2.52	
6 1 -1 1 -1 2.48 2.55	
7 1 -1 -1 1 8.12 7.98	
8 1 -1 -1 -1 4.15 4.26	
9 -1 1 1 1 0.72 0.86	
10 -1 1 1 -1 1.27 1.36	
11 -1 1 -1 1 1.18 1.13	
12 -1 1 -1 -1 2.62 2.52	
13 -1 -1 1 1 0.71 0.67	
14 -1 -1 1 -1 1.48 1.35	
15 -1 -1 -1 1 1.38 1.45	
16 -1 -1 -1 -1 3.31 3.25	
run;	Start data input
Proc Glm;	Start regression procedure
Model y =x1 x2 x3 x4 x1*x2 x1*x3 x1*	Define the model form
x4 x2* x3 x2*x4 x3*x4;	
run;	Start regression calculation

macro would suggest a model as follows:

$$Y = 2.99 + 1.41 \times X_1 - 1.04 \times X_3 + 0.3 \times X_4 - 0.51 \times X_1 \times X_3$$
$$+ 0.87 \times X_1 \times X_4 - 0.31 \times X_3 \times X_4$$

The size of the coefficients corresponds to the contribution level of each term, because the model is built through regression with a coded experimental data set.

3.6.3.2 SYSTAT for Windows™

The operation of this program in analyzing regression is quite simple. One can type the actual experimental data into a spreadsheet or import them with the

TABLE 3.15. Regression Result of SAS Macro in Table 3.14.

SAS   12:22 Wednesday, May 1, 1996 1
               General Linear Models Procedure

Dependent Variable: Y

Source	DF	Sum of Squares	Mean Square	F Value	Pr > F
Model	10	138.4417812	13.8441781	27.92	0.0001
Error	21	10.4130406	0.4958591		
Corrected Total	31	148.8548219			

	R-Square	C.V.	Root MSE	Y Mean
	0.930046	23.53862	0.704173	2.99156250

SAS   12:22 Wednesday, May 1, 1996 3
               General Linear Models Procedure

Dependent Variable: Y

Source	DF	Type I SS	Mean Square	F Value	Pr > F
X1	1	63.87325312	63.87325312	128.81	0.0001
X2	1	0.04727813	0.04727813	0.10	0.7605
X3	1	34.42425312	34.42425312	69.42	0.0001
X4	1	2.92215313	2.92215313	5.89	0.0243
X1*X2	1	0.21945313	0.21945313	0.44	0.5131
X1*X3	1	8.35382813	8.35382813	16.85	0.0005
X1*X4	1	24.13387812	24.13387812	48.67	0.0001
X2*X3	1	0.92820312	0.92820312	1.87	0.1857
X2*X4	1	0.50752813	0.50752813	1.02	0.3232
X3*X4	1	3.03195312	3.03195312	6.11	0.0220

SAS   12:22 Wednesday, May 1, 1996 5
               General Linear Models Procedure

Dependent Variable: Y

Parameter	Estimate	T for H0: Parameter=0	Pr > |T|	Std Error of Estimate
Intercept	2.991562500	24.03	0.0001	0.12448131
X1	1.412812500	11.35	0.0001	0.12448131
X2	-0.038437500	-0.31	0.7605	0.12448131
X3	-1.037187500	-8.33	0.0001	0.12448131
X4	0.302187500	2.43	0.0243	0.12448131
X1*X2	0.082812500	0.67	0.5131	0.12448131
X1*X3	-0.510937500	-4.10	0.0005	0.12448131
X1*X4	0.868437500	6.98	0.0001	0.12448131
X2*X3	0.170312500	1.37	0.1857	0.12448131
X2*X4	0.125937500	1.01	0.3232	0.12448131
X3*X4	-0.307812500	-2.47	0.0220	0.12448131

FILE/OPEN option in the menu and then choose the regression icon from the toolbar in the MAIN window (Figure 3.13). Sometimes a macro (Table 3.16) can be used with the NOTEPAD window to get more information about regression analysis.

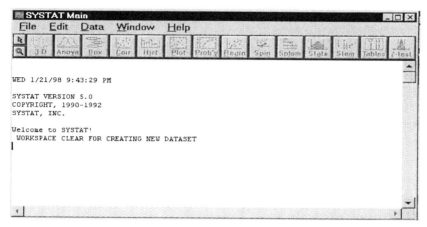

Figure 3.13 The start user interface of the SYSTAT.

3.6.3.3 STATISTICA

STATISTICA is one of the most powerful and widespread statistical packages containing all kinds of statistical procedures, from basic to industrial statistics, including quality control, process control, and experimental design. Its Experimental Design module has extensive functions that can generate different kinds of traditional designs (Figure 3.14). The user interface of the Microsoft Windows® compatible version is quite user-friendly. Most statistical operations can be finished through mouse clicks.

3.6.3.4 SigmaStat

SigmaStat is a statistical software package from Jandel Scientific, Inc., which meanwhile has been purchased by SPSS. It supplies forward and backward

TABLE 3.16. **Regression Macro for SYSTAT.**

Macro	Explanation
Use 'Hu_reg01.SYS'	Input the variables and data from file "hu_reg01.sys"
MGLH	Start the multiple regression procedure
Model Y = constant + X1 + X2 + X3 + X4 + X1*X2 + X1*X3 + X1*X4 + X2*X3 + X2*X4 + X3*X4	Define the model form
Estimate	Start the regression calculation
Corr	Calculate the correlation matrix between all variables
Pearson X1,X2,X3,X4, Y / Prob, listwise	

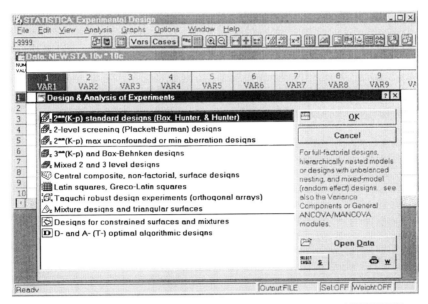

Figure 3.14 The start user interface of the experimental design procedure of STATISTICA.

multiple regression procedures. However, it can only analyze linear multiple regression directly based on the data supplied. If a quadratic model [Equation (3.7)] is intended to be build, a manual treatment of data for the estimate of the quadratic terms is required. The nonlinear regression procedure may be used for estimation of a quadratic model.

3.6.3.5 StatGraphics

This program has, besides the normal regression procedure, an experimental plan design module that contains a series of classical statistical experimental designs. The modeling and analysis based on the experimental data are relatively simple. However, the exclusion of drop-off terms from the model can only be performed manually.

3.6.3.6 MODDE

MODDE is an extensive program specialized for experimental design and data analysis. It contains almost all necessary functions related to product and process optimization in terms of statistical experimental design. MODDE is useful for the layman who does not want to make the effort to understand statistics. Figure 3.15 is the typical user interface of MODDE.

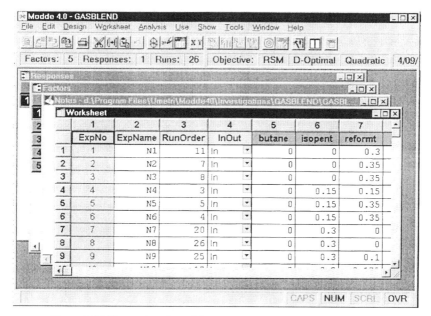

Figure 3.15 The user interface of the experimental design program MODDE.

3.6.4 MODEL VALIDITY TEST

A significant and precise model contributes substantially to the development of an organized set of information concerning the relationship between food product characteristics and processing conditions. Before a model is used for further application, its significance or precision must be tested. Although the terms in the model have been estimated in such a way that the equation can optimally predict the experimental data through regression, some statistical criteria based on the residuals are used to judge how well a model describes the experimental data.

The P value of an F-test run on the model and the R^2 and adjusted R^2 values are general statistical diagnostic testers. They are based on variance analysis and normally used to judge the adequacy and quality of the model fit and the experimental error. They are discussed in more details in the following section.

If a model proves unsatisfactory, the food researcher has to either perform some kind of mathematical transformation of the variables or add more supporting data points and change the form of the assumed model to improve the model fit. The Box-Cox transformation can be used to get essential information of how to transform a model effectively.

3.6.4.1 *F*-Test

The *F*-test is the most frequently used statistical test to check the adequacy or significance level of a model. It is based on the following statistical hypothesis: the description of the experimental data by their average value is significantly better than the predicted model value. In this test, a comparison between the predicted values by the model and the actual experimental data is performed. The *F*-value for the model is calculated by dividing regression mean squares by residual mean squares [Equation (3.18)], and the significance level (*p*-value) of the model is found by comparing the *F*-value with the standard *F*-value in a statistical *F*-value table. With a good model fit, the variance around the regression curve is significantly lower than the variance around the average. The degree of freedom of the regression (DF_{reg}) is the number of regression coefficients minus one, and that of the residual (DF_{rest}) is the number of observations or trials subtracted by the number of regression coefficients. Mean squares (*MS*) are calculated by dividing the sum of squares by the corresponding degree of freedom.

$$F = \frac{\text{Regression mean squares}}{\text{Residual mean squares}}$$

$$= \frac{MS_{reg}}{MS_{rest}} \tag{3.18}$$

where

$$MS_{reg} = SS_{reg}/DF_{reg} \qquad \text{mean of explained sum of squares (\textit{SS})}$$
$$MS_{rest} = SS_{rest}/DF_{rest} \qquad \text{mean of residual sum of squares}$$

$$SS_{reg} = \sum_{i=1}^{m} \left(Y_{(X_i)} - \overline{Y} \right)^2 \qquad \text{explained sum of squares: squared sum of deviations between predicted function values } Y_{(Xi)} \text{ and the average dependent variable value } \overline{Y}$$

$$SS_{rest} = \sum_{i=1}^{m} \left(Y_i - Y_{(X_i)} \right)^2 \qquad \text{residual sum of squares: squared sum of deviations between measured experimental values } Y_i \text{ and predicted dependent variable values } Y_{(X_i)}.$$

$Y_{(x_i)}$: Predicted value of dependent variable at $X = X_i$
Y_i : Real response value at $X = X_i$
$\overline{Y}$: Average value of response of all trials
m : Trial number

The *F*-test gives a probability *p*, which indicates the level of significance. The *F*-value is calculated according to Equation (3.18) and is compared with the critical table value of $F^{\alpha}_{(FGreg, FGrest)}$. The hypothesis is rejected at the *p*-level with the probability of $1 - p$, if the value of the *F*-value exceeds the critical table value. In other words, the significance *p* shows the statistical risk level for rejecting the assumption that the experimental data can be predicted significantly better by the average value than by the model. Models are often described according to their risk level ranges with the words "almost significant" $(0.05 > p \geq 0.10)$, "significant" $(0.05 \geq p > 0.01)$, and "very significant" $(0.01 \geq p)$. In the industrial engineering field, a significance level *p* of 0.05 or even 0.10 according to the *F*-test may be adequate.

It must be noted that the *F*-value is extremely sensitive to the number of degrees of freedom for experimental error. In general, the degrees of freedom for error estimation should be more than three to five. For example, a 2^n factorial design with less than four variables supplies only a small number of degrees of freedom. It is suggested that repeating some or even all the trials in the design may get more degrees of freedom. For an experimental design of the 2^n type with more than four variables, the number of degrees of freedom increases rapidly with the trial number. Therefore, a replication of the trials is not always necessary. If no trials are repeated, the experimental variance of the experimental data can, in principle, be determined from the insignificant multifactor interactions between the variables.

3.6.4.2 Measures for Model Fit and Predictive Power

Besides the *F*-test, another important criterion to examine the quality of the model is the multiple correlation coefficient *R* (multiple *R* in SPSS), which describes the linear correlation between the independent Y_i and its predicted value $Y_{(Xi)}$. A large value of *R* normally indicates a workable regression equation. What may appear even more frequently in the literature is the coefficient of determination R^2, which indicates the goodness of fit. It is a percentage ratio of the through model explained sum of square (SS_{reg}) to the total sum of square (total variance SS_{total}). In other words, the extent of the variance can be explained with the model [Equation (3.19)]. The total variance is also known as the "should be explained sum of squares."

$$R^2 = \frac{\text{Explained sum of squares}}{\text{Total sum of squares}}$$

$$= \frac{SS_{reg}}{SS_{total}} \tag{3.19}$$

where

$$SS_{total} = \sum_{i=1}^{m}(Y_i - \overline{Y})^2 = SS_{reg} + SS_{rest}$$

R^2 is a value between 0 and 1. A value of 1 means that the model perfectly predicts all of the experimental data and a values of 0 means totally faulty prediction (the average value of the data is an equal or better predictor than the model). In practical application, R^2 values smaller than 0.75 usually indicate an insufficiently precise description of the experimental data by the model. Obviously, if the R^2 value does not reach statistical significance, or if it is quite small, say 0.50, proceeding to the next stage of food product optimization is certainly not appropriate. In addition, it must be kept in mind that R^2 is a ratio. Sometimes, a big value of R^2 cannot mean that the absolute residuals are small.

Actually, R^2 is an overestimate of goodness of fit. It is usually modified into an adjusted R^2 by taking into account of the degrees of freedom [Equation (3.20)]. Adjusted R^2 is a more suitable measure for judging the goodness of fit of a model.

$$R^2_{\text{Adj}} = \frac{MS_{\text{total}} - MS_{\text{rest}}}{MS_{\text{total}}} \qquad (3.20)$$

The predictive power of a model is given by Q^2 which is based on the prediction residual sum of squares (PRESS). This is a measure of how well the model will predict the responses for new experimental conditions. PRESS is computed as the squared differences between observed Y_i and predicted values $Y_{(Xi)}$ and is computed as follows [Equation (3.21)]:

$$Q^2 = \frac{MS_{\text{total}} - \text{PRESS}}{MS_{\text{total}}} \qquad (3.21)$$

A Q^2 value larger than zero indicates that the dimension is significant (predictive). Large Q^2 value, say 0.7 or larger, indicates that the model has good predictive ability and will have small prediction errors. In contrast to R^2, Q^2 is an underestimate of goodness of fit of a model.

3.6.4.3 Sources of Errors

The model built is an empirical function that can never describe the experimental data with 100% accuracy because of unavoidable errors that occur during experimentation and analysis. The sensory, physical, or chemical analytical methods are always accompanied by random experimental or system errors. These errors can and should be controlled and kept small enough that the significance of independent variables for the food quality indices of interest can be recognized. In statistical food product design approaches the experimental error is usually checked and determined by replication of trials.

Other causes for inadequate modeling might be the unsuitable assumptions made in constructing the model. In Figure 3.16 the sources of variance are analyzed and classified. In a word, an unexplained variance is normally caused by model weakness, the system and random experimental errors. In Figure 3.17

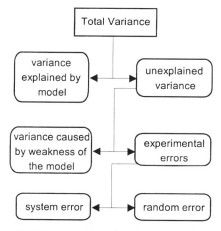

Figure 3.16 Decomposition of total experimental variance.

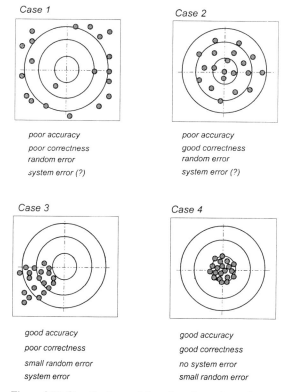

Figure 3.17 Classification and detection of experimental error.

the basic concept of system and random errors is demonstrated with the example of a target-shooting practice. In cases 1 and 2 the total error may consist of mere random and system errors, whereas in case 3 an obvious system error exists.

It must be mentioned that, in some cases, faulty documentation of experimental data (faulty trial data registration and transfer) and imprecise computer programs are the real causes of inadequate or unsuccessful modeling. Faulty data usually lie far away from the whole data set and can be easily identified. These data in the original documentation must be rechecked and should not be deleted or changed without enough evidence. The system data failure can be recognized by a usual pattern of regular and rhythmical deviations. Only use tested and proven software packages. Never assume that newly obtained software can do a good job with high accuracy. Check the software and make sure it is correct before using it in food product design.

3.6.4.4 Model Test with Software

The regression procedure of most standard software packages, such as STATISTICA, SPSS, and SAS, supplies information of model tests as well as the normal estimates of coefficients. The statistical model tests are usually part of the standard regression analysis output results. A food product developer usually does not need to concern himself about the process of calculation of the model's diagnostic criteria but only about the decision whether the model built is good enough according to the specified criteria.

3.6.5 A PRACTICAL EXAMPLE

As a practical example, this section presents a study on how to inactivate the trypsin inhibitor (TI) in soybeans through extrusion cooking. Three important operating variables, namely, the mass temperature $T\,(^{\circ}C)$, mass moisture content $M(\%)$, and screw rotating speed $n\ (\text{min}^{-1})$, were controlled, and their effect on degree of inactivation of TI $D_{TI}\ (\%)$ was studied. The experiments were performed according to a CCD with three repeated trials at the central point. The α value for the calculation of the star point is 1.4. The selected test levels of each variable are listed in Table 3.17.

TABLE 3.17. Selected Test Levels of Each Extrusion Variables.

Level Variable	Lower Star Point	Lower Level	Middle Level	Upper Level	Upper Star Point
Code	$-\alpha$	-1	0	$+1$	$+\alpha$
T	123	130	150	170	177
M	16	18	24	30	32
N	113	120	140	160	167

TABLE 3.18. Designed and Real Experimental Data.

	Designed Experimental Plan			Actual Real Experimental Data			
No	T	M	n	T	M	n	D_{TI}
1	130	120	18	126	18	118	82.65
2	130	120	30	130	18	166	81.19
3	130	160	18	129	30	124	77.83
4	130	160	30	131.5	30	160	94.61
5	170	120	18	158	18	120	91.90
6	170	120	30	161	18	160	95.39
7	170	160	18	164	30	120	84.56
8	170	160	30	158	30	165	90.09
9	123	140	24	122	24	143	58.81
10	177	140	24	170	24	140	97.77
11	150	113	24	150	24	113	80.80
12	150	167	24	149	24	171	88.80
13	150	140	16	150	16	140	93.80
14	150	140	32	150	32	139	81.90
15	150	140	24	151	24	140	87.72
16	150	140	24	150	24	140	87.91
17	150	140	24	151	24	140	88.05

According to these levels, a CCD for three variables was set up and performed. The test levels of variables cannot always be controlled at the levels intended or designed in the practical trial. The designed test levels and their corresponding actual values are recorded and listed in Table 3.18. The true experimental data were saved in ASCII format in a separate file "S_TIU01.dat".

A quadratic model was intended to be developed to fit these real experimental data. The software package SPSS of PC version SPSS/PC+ was used, and the content of the macro using backward stepwise regression procedure was saved under the file "s_tiu01.inc" as follows:

```
data list file='a:\ dat\ s_tiu01.dat'free/
TM n DTI.
compute TM = T*M.
compute Tn = T*n.
compute Mn = M*n.
compute TT = T*T.
compute MM = M*M.
compute nn = n*n.
set listing = 'a:\ lis\ s_tiu01.lis'.
regression variables = T to nn/dependent = DTI/method = backward/ casewise = all.
correlation variables = T to DTI.
finish.
```

To start the regression procedure with the macro, the SPSS/PC+ working environment was started with the command SPSSPC behind a DOS prompt.

Then, in the SPSS environment, the command "INCLUDE s_tiu01.inc" was entered to start the macro. The macro used the data from the file "s_tiu01.dat", and the regression results were shown on the screen and written into the file "s_tiu01.lis", as defined in the macro. The output file "s_tiu01.lis" can be viewed with word processors, such as Ms-Word®, Ms-Editor®, Lotus AmiPro®, or WordPerfect®. The most important part of this file is printed in Table 3.19.

TABLE 3.19. Regression Results: Content of File s_tiu01.lis (part).

```
The raw data or transformation pass is proceeding
   17 cases are written to the uncompressed active file.
-------------------------------------------------------------------------
Page 2                    SPSS/PC+                    3/26/96
                    {Some part here is omitted}
              **** MULTIPLE REGRESSION ****
Equation Number 1   Dependent Variable.. DTI
Variable(s) Removed on Step Number
 7.. MM
Multiple R    .74643
R Square      .55715
Adjusted R Square    .45496
Standard Error      6.70009
Analysis of Variance
              DF    Sum of Squares    Mean Square
Regression    3      734.21365        244.73788
Residual     13      583.58534         44.89118
F = 5.45180        Signif F = .0120
-------------------------------------------------------------------------
Page 11                   SPSS/PC+                    3/26/96
              **** MULTIPLE REGRESSION ****
Equation Number 1   Dependent Variable.. DTI
--------------Variables in the Equation---------------------
```

Variable	B	SE B	Beta	T	Sig T
TM	-8.57835E-03	4.24717E-03	-.78910	-2.020	.0645
TT	2.177923E-03	5.51850E-04	.98921	3.947	.0017
MN	6.378934E-03	3.74713E-03	.59664	1.702	.1125
(Constant)	47.23454	11.84990	3.986	.0016	

```
-------------------------------------------------------------------------
Page 12                   SPSS/PC+                    3/26/96
-------------Variables not in the Equation--------------------------
```

Variable	Beta In	Partial	Min Toler	T	Sig T
T	3.86680	.25968	1.9795E-03	.932	.3700
M	1.14109	.15168	6.4902E-03	.532	.6047
N	-.91141	-.33684	.01679	-1.239	.2389
TN	-.94881	-.28055	.01613	-1.013	.3313
MM	1.88032	.35404	.01267	1.311	.2143
NN	-.77980	-.30218	.01847	-1.098	.2937

```
-------------------------------------------------------------------------
```

(continued)

TABLE 3.19. (Continued)

```
-------------------------------------------------------------------------
Page 13                    SPSS/PC+                    3/26/96
                    {Some part here is omitted}
          ****  M U L T I P L E   R E G R E S S I O N  ****
Equation Number 2   Dependent Variable.. DTI
Casewise Plot of Standardized Residual *: Selected  M: Missing
-3.0   0.0   3.0
 Case #      O:............:O        DTI          *PRED          *RESID

    1   .          . *     .        82.65         77.6073         5.0427
    2   .          .*      .        81.19         79.0652         2.1248
    3   .          *       .        77.83         78.6964         -.8664
    4   .          . *     .        94.60         79.6236        14.9764
    5   .          .*      .        91.90         90.5460         1.3540
    6   .          .*      .        95.39         91.9085         3.4815
    7   .      *   .       .        84.56         93.2966        -8.7366
    8   .        * .       .        90.09         90.5460         -.4560
    9   .   *      .       .        58.81         76.1950       -17.3850
   10   .          .*      .        97.77         96.1497         1.6203
   11   .       *  .       .        80.80         87.0380        -6.2380
   12   .         .*       .        88.80         86.6123         2.1877
   13   .          . *     .        93.80         87.0380         6.7620
   14   .        * .       .        81.90         87.0380        -5.1380
   15   .          *       .        87.72         87.4665          .2535
   16   .          *       .        87.90         87.4665          .4335
   17   .          *       .        88.05         87.4665          .5835

-------------------------------------------------------------------------
Page 21                    SPSS/PC+                    3/26/96
          ****  M U L T I P L E   R E G R E S S I O N  ****
Equation Number 3   Dependent Variable.. DTI
Residuals Statistics:
                  Min        Max        Mean      Std Dev    N

*PRED           76.1950    96.1497     86.1035     5.8686    17
*RESID         -17.3850    14.9764     -.0000      6.9226    17
*ZPRED          -1.6884     1.7119     -.0000      1.0000    17
*ZRESID         -2.4316     2.0947     -.0000       .9682    17
 Total Cases = 17

-------------------------------------------------------------------------
Page 23                    SPSS/PC+                    3/26/96
Correlations:        T          M          N          DTI

T                 1.0000      .0387     -.0257       .6552*
M                  .0387     1.0000      .0148      -.1614
N                 -.0257      .0148     1.0000       .2266
DTI                .6552*    -.1614      .2266      1.0000
N of cases: 17       1-tailed Signif: * -.01 ** -.001
"." is printed if a coefficient cannot be computed

-------------------------------------------------------------------------
Page 24                    SPSS/PC+                    3/26/96
This procedure was completed at 18:46:15
End of Include file.
```

85

The full model involves a constant and the terms of $T \cdot M$, $T \cdot n$, $M \cdot n$, T^2, M^2, and n^2. However, a simpler model of the third equation, obtained from the backward regression procedure, with fewer terms of constant, $T \cdot M$, T^2, and $M \cdot n$, was selected for practical application [Table 3.19, Equation (3.22)]. Please note that this model is based on the regression of the actual experimental data (not coded) and so that only real variables values can be used in this model. The statistical diagnostic criteria, namely, the significant level p and the multiple R were about 0.01 and 0.75, respectively.

$$D_{TI} = 47.23 + 2.18 \times 10^{-3} \times T^2 - 8.58 \times 10^{-3} \times T \times M$$
$$+ 6.38 \times 10^{-3} \times M \times n \tag{3.22}$$

3.7 EFFECT ANALYSIS OF PROCESS VARIABLES

Model building is not the original intent of statistical food product design. The interests of food developers focus on the effects of various factors on the food quality, namely, which variables in which ranges have relevant effects on defined food quality attributes. In other words, under which conditions can foods with defined quality indices be produced. The food product researcher should outline the advantages of the optimum food product, provide information about its production control, and warn of its difficulties and potential problems during production.

Different food quality indices might be affected by independent variables in a different manner. A significant variable for food quality A may not be important for quality index B, and thus two models, which sometimes may include a different set of terms, must be applied to analyze the factor effects on A and B.

As discussed above, only a really significant and precise model can supply reliable essential information for food developers. To extract this information from a valid model obtained, two approaches are generally used. One is the graphical approach, in which the effect of each variable or of multiple variables on different food qualities are plotted in graphs such as cube plots or response surfaces. This helps to visually discern relationships between factors and responses and may point out regions of interest for further investigation. The other approach is a numerical method, which may supply more exact information about the optimum. However, the numerical information is usually not as easily interpreted.

3.7.1 GRAPHICAL APPROACHES

In the graphical approach, the predictive models are used to generate contours, response surfaces, or even contour surfaces within the experimental region with the help of suitable computer software. The contour plots can be created in two- or three-dimensional spaces, respectively. They represent information for two to three factors and one or more responses. The effect of these included variables on the food quality indices can be judged by visual inspection of the contours or surfaces. The surfaces and contour plots are reasonably accurate within the experimental region and can be used as guidelines during food processing, quality management, and product development.

The graphical method for effect analysis of each variable reduces drastically the time and effort required. It also provides the food researcher with an informative insight into the system as well as an understanding of how the system might behave when the levels of factors are changed. Additionally, these plots might reveal the interrelationships among the test variables, describe the combined effect of all test variables on the response, indicate optimum regions for achieving specific food product attributes, and lead to rapid process optimization.

In general, two-dimensional (2D) X-Y plots, three-dimensional (3D) response surfaces, 2D contour plots, 3D + 2D graphics and 3D contour surface charts are useful to represent the effects of one or more factors on the food quality indices of interest. Different graphics are actually a simulation of the real relationship between the variables and the response and can be understood easily. A model may be described with different kinds of graphics, but the essential characteristics revealed by them are the same. The different graphic-generating techniques are shown step by step in the following sections.

3.7.1.1 2D X-Y Plot

The effect of a single independent variable on the target parameter—a food quality index—can be plotted in a traditional X-Y plot. For example, to show the single effect of mass temperature T alone on the inactivation of trypsine inhibitor D_{TI} [%], the model 3.20 was used by fixing the mass moisture content M and the screw rotation speed n at constant levels, say, at four different extreme levels (16%, 113 min^{-1}), (16%, 167 min^{-1}), (32%, 113 min^{-1}), and (32%, 167 min^{-1}). To draw these four curves or lines in a T-D_{TI} two-dimensional coordinate system, a so-called table calculation (spreadsheet) computer program, such as Ms-Excel® and Lotus 1-2-3®, is usually helpful. The following shows how to plot these four curves in three steps:

3.7.1.1.1 Step I

Type the tested temperature levels from 123 to 177°C in a step of 6°C (the step can be still smaller) in column A in a spreadsheet of Ms-Excel®. Select the four columns B, C, D, and E for the storage of D_{TI} data to be calculated for the four curves to be drawn. Move the cursor to cell B3 and type the equal mark

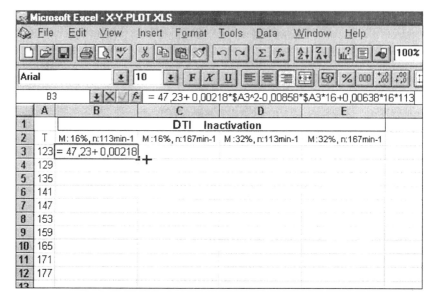

Figure 3.18 Step I of generating an X-Y plot.

"=" and then the model behind it, in which T, M, and n should be rewritten as $A3$, 16, and 113, respectively (Figure 3.18).

3.7.1.1.2 Step II

Move the mouse arrow ($\mathbb{\diagdown}$) to the right-bottom corner of cell B3. The arrow will turn into the symbol +, drag it to cell E3 with left mouse key pressed. Select the cells C3, D3, and E3, and rewrite the M and n value (here 16 and 113) with the values in the upper cell C2, D2, and E2, respectively. Select cell B3 to E3 with the mouse, and drag them all the way to row12 with left mouse key pressed. Now the data for these four curves are calculated one at a time (Figure 3.19).

3.7.1.1.3 Step III

Select all the data from cell A2 to E12 with the mouse. Click the button ![button] (ChartWizard) from the toolbar with the mouse and then drag a box on the spreadsheet with the left mouse key pressed. A new window will pop up for further graphic operation. Choose the curve symbol (marked black), and click the button ![button] in the popped window (Figure 3.20). Finally, the finished graphic will be generated as in Figure 3.21.

All the four curves in the figure are almost straight lines, which means that the effect of temperature on the D_{TI} is almost linear. The D_{TI} shown is partly an extrapolation, because it is impossible to achieve a TI inactivation of more than

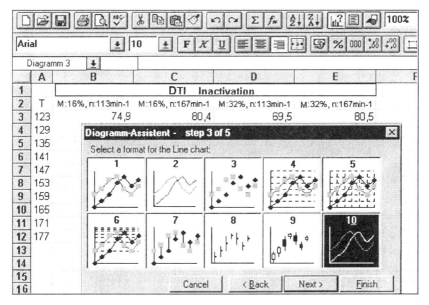

Formula bar: B3 | = 47,23 + 0,00218*$A3^2 - 0,00858*$A3*16 + 0,00638*16*113

T	M:16%, n:113min-1	M:16%, n:167min-1	M:32%, n:113min-1	M:32%, n:167min-1
	DTI Inactivation			
123	74,9	80,4	69,5	80,5
129	77,3	82,8	71,2	82,2
135	80,0	85,5	73,0	84,0
141	82,7	88,3	74,9	86,0
147	85,7	91,2	77,0	88,1
153	88,8	94,3	79,3	90,3
159	92,1	97,6	81,8	92,8
165	95,5	101,0	84,3	95,4
171	99,0	104,5	87,1	98,1
177	102,8	108,3	90,0	101,0

Figure 3.19 Step II of generating an X-Y plot.

100%. The slope of all four curves in Figure 3.21 reveals that the inactivation of TI increases with increased temperature from 123 to 177°C. However, the temperature affects the D_{TI} more intensively at low mass moisture content (compare the slope of lines —X— and —Δ— with —◇— and —○—). The highest D_{TI} was achieved at low mass moisture content but high screw rotation

Figure 3.20 Step III of generating an X-Y plot.

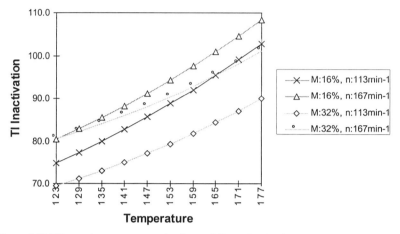

Figure 3.21 Effects of temperature on the D_{TI} at different levels of mass moisture *m* and screw rotation speed *n*.

speed (line —△—), whereas the lowest D_{TI} was reached at high mass moisture content but low screw rotation speed and low temperature (line —◇—).

3.7.1.2 3D Response Surface

For a better visualization, the variable effects on a definite response can be plotted in a 3D space, usually in the form of a fishnet surface with/without color gradient corresponding to the size of the response value. The model is illustrated in a 3D surface in which two factors are presented on horizontally perpendicular axes and the response on the vertical axis. Such a surface is known as a response surface. If more than two independent variables are involved in the model, all other variables except the two selected must be fixed at certain, usually their middle, levels. Thus, the original model is simplified to an equation with only two variables for 3D response surface plotting. It is reasonable to select these two most important variables to generate the 3D plot so that their effects are presented in the diagram. However, sometimes a minor but meaningful factor and a significant one are chosen for plotting to stress the obvious effect difference between them. The food researcher should remember that the variation ranges of these two selected variables should lie only within the researched region, because any large extrapolation may cause inexact and even incorrect results.

A response surface is actually a surface based on the assumption that all the independent variables are continuous in the experimental area. Response surfaces are not always easy to interpret because they can be sometimes in a wide variety of shapes, such as plane, cradle, bowl, saddle, and bell. A straight plane means that the significant effects of these two variables on the response are merely linear, whereas a bent plane shows an additional relevant interaction of these two variables on the response. The more the plane bends, the more significant is the interaction between the two variables. Large curvature effects, represented by terms, such as X_j^2, produce parabolic shapes when the model is graphed. They occur when two different levels of the same factor produce similar values of response but higher or lower responses at the intermediate test level. Generally, the response surface can be read and understood like a topographical map. For example, by moving along the X-axis, the negative or positive effect of the variable X is revealed, depending on the downward or upward tendency of the surface.

In Figure 3.22 the Model 3.20 is plotted by fixing the screw rotation speed n at its middle level 140 min^{-1} to check how the mass temperature and mass moisture content affect the inactivation of trypsine inhibitor. In this way Model 3.20 is simplified into Equation (3.23), and the response surface would be generated accordingly:

$$D_{\text{TI}} = 47.23 + 2.18 \times 10^{-3} \times T^2 + 0.8932 \times M - 8.58 \times 10^{-3} \times T \times M$$

$$(3.23)$$

From the graphic, the effect of mass temperature T and moisture content M can be visually examined. It is clear that the mass temperature has the most significant effect on the TI inactivation, whereas the moisture content plays only a minor role. Increasing mass temperature while decreasing mass moisture

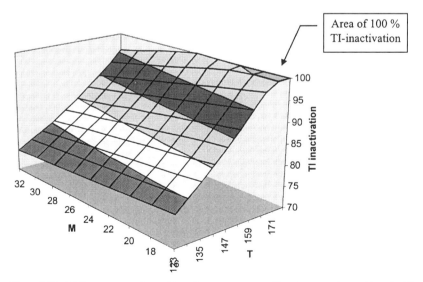

Figure 3.22 A 3D response surface: effect of mass temperature T and mass moisture M on the TI inactivation (%).

content leads to a higher degree of TI inactivation. The most effective—even completely—TI inactivation will be reached at mass temperatures higher than 171°C and mass moistures lower than 18%.

Numerous software packages can be used to generate 3D response surface plots. Some of them are listed at the end of this section. To get a 3D plot, one merely needs to input the ranges of these two independent variables selected, and the plot will be generated by the computer automatically. Usually, this response surface can be rotated to get a clearer perspective. Sometimes, it is important to set the Z-axis (response) value in a meaningful range of interest for a realistic exploration of the graphic.

A statistical software package is usually expensive. To overcome this problem, some popular table calculation programs, such as Ms-Excel® and Lotus 1-2-3®, can also be used to generate surface plots. In the following the techniques of creating the surface plot using Equation (3.21) are presented step by step.

3.7.1.2.1 Step I

Choose and type values for the mass temperature T (123–177°C with a step of 6°C) and the moisture content M (16–32% with a step of 2%) in column A and row 1, respectively. Move the cursor to cell B2; type "=" and then type the Equation (3.23) in which T and M are replaced with $A2 and B$1 (Figure 3.23). The symbol "$" in front of A2 ($A2) but between B and 1 (B$1) ensures that the T value would only be changed when the row number changes, the M value varies

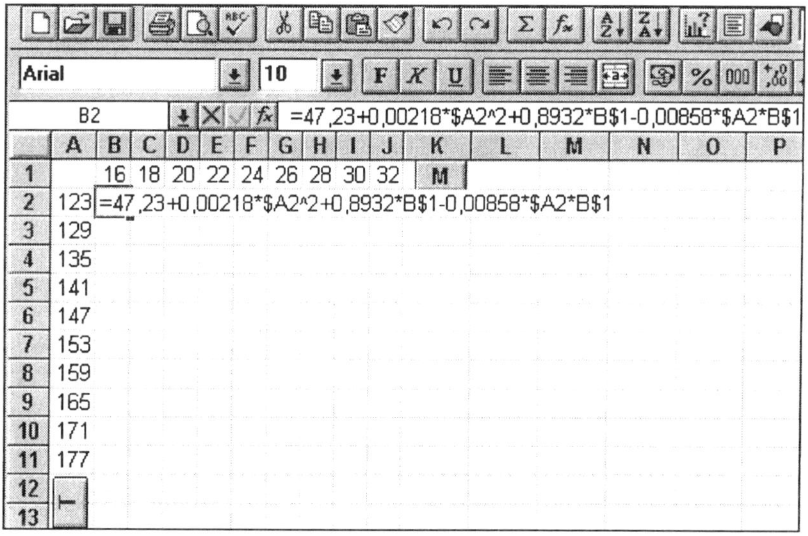

Figure 3.23 Step I of generating a 3D response surface plot using Ms-Excel®: model input.

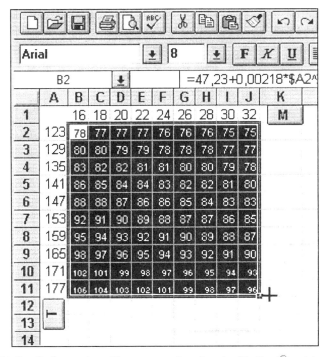

Figure 3.24 Step II of generating a 3D response surface plot using Ms-Excel®: matrix calculation.

only if the column number varies. Press the ENTER key and the number "78" will show in cell B2 ([78] surrounded by a dark frame) instead of the equation.

3.7.1.2.2 Step II

Move the mouse cursor (symbol: ↘) to the bottom-right corner of cell B2 and drag it to cell B11 with the pressed left mouse key; then drag cell B11 further to J11 to calculate all the response values as shown (Figure 3.24). Each response value in the region marked black is actually calculated according to the model with corresponding T and M levels, which locate cross-linked to the cell. Now a set of data will be supplied in the form of a T-M-D_{TI} matrix that contains T values in the column A, M values across the row 1, and D_{TI} values filling in the spreadsheet matrix.

3.7.1.2.3 Step III

With the left mouse taste pressed, move the mouse arrow to cell A1 and drag it to cell J11 to select the whole data matrix (within the black frame in Figure 3.25). Select the button 📊 from the toolbar (or INSERT then CHART

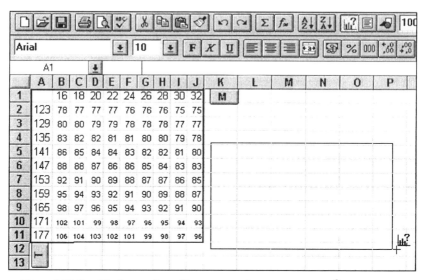

Figure 3.25 Step III of generating a 3D response surface using Ms-Excel®: start plotting.

form the menu). The mouse arrow will now change into the symbol 𝖑𝖚?. Again, with the left mouse key pressed, drag the mouse cursor to form a frame on the worksheet in which the surface graphic will be drawn.

3.7.1.2.4 Step IV

Lose the left mouse taste and a new interactive window will pop up. Click the button ▦ⁿᵉˣᵗ▦ with the left mouse key to go on, and select the graphic type of "3D surface" to generate the 3D surface plot (Figure 3.26).

3.7.1.2.5 Step V

Select the first column as X-axis label and the first row as Y-axis label. Click on the button ▦ⁿᵉˣᵗ▦ to go on and follow the command in the interactive window (Figure 3.27). The 3D surface of the given model will be plotted in the frame on the worksheet (Figure 3.28). This 3D graphic can be easily reprocessed with the mouse to rotate to a clearer perspective. The axes scales can be changed, and the color of the background can be modified. A quick double-click of the left mouse key on any part of the chart or a click of the right mouse key will lead to pop-up of interactive windows or menu in which the properties of the graphics can be modified. This graphic can be added to any Ms-Windows® applications through the COPY and PASTE commands. This is one of the most common advantages of Ms-Windows®. Figure 3.22 is obtained through copying this diagram in Ms-Excel® and adding it in Ms-Word® for Windows®.

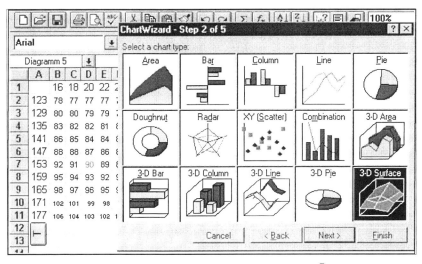

Figure 3.26 Step IV of generating a 3D response surface using Ms-Excel®: select the chart type.

In food product design there is usually more than one food quality index to be considered at the same time. Therefore, more than one response surface of different responses can be combined on the same 3D diagram for easy comparison. It must be noted that the two selected independent variables and the ranges in which they vary must be the same. One can draw these two 3D graphics

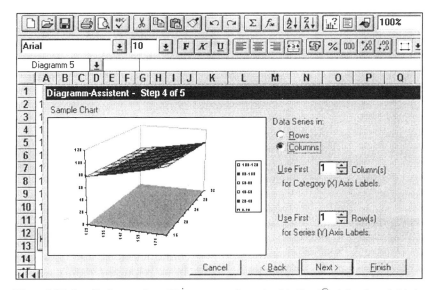

Figure 3.27 Step V of generating a 3D response surface using Ms-Excel®: define the axis labels.

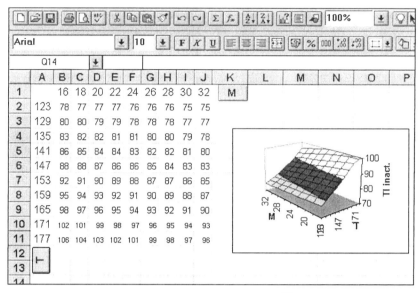

Figure 3.28 3D response surface generated using Ms-Excel®: create the surface plot.

separately and then overlay them simply in a Draw or CAD program such as Micrograx Charisma®, Micrograx Designer®, or Corel Draw®. Taking advantage of the features of overlaid surfaces, the effect of variables on different responses can be directly interpreted.

3.7.1.3 2D Contour Plot

The information presented in a 3D response surface diagram can be presented in a 2D plot. In a 2D contour plot, there are several contour lines representing a series of response values. Geometrically, these contour curves are projections of cross sections of the 3D response surface onto the plane of the two variables. The heights of the cutting planes are corresponding to the response value of the related contour curves. Each contour represents a certain response value and shows the factor levels responsible for that response. A fixed interval of the response value is usually used to generate contour lines for easy analysis of the variable effect in different research regions.

The technical foundation of creating a 2D contour is similar to that of generating a 3D response surface. The computer programs used for generating surface charts also support normally the feasibility of creating 2D contours plots. Figure 3.29 shows the contour plot from Model 3.21 generated with Ms-Excel®. Its plotting steps are similar to those taken for creating a response surface as described above. The only difference is that a contour graphic must be selected instead of a 3D graphic after step IV (no. 3 instead of no. 1 in Figure 3.30). A double-click of the mouse or a single-click of the right mouse key on the chart

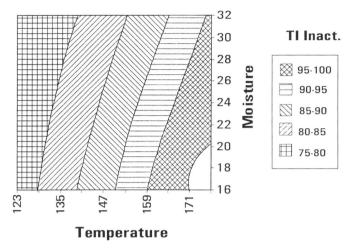

Figure 3.29 2D contour: effects of temperature and moisture content on TI inactivation.

will open an interactive window or menu in which the contour properties, such as the number of contour lines, can be modified.

Sometimes a 2D contour plot provides the researchers with more exact information about the relationship between the variables and the food quality indices (responses). In Figure 3.29 the whole surface is divided into six regions that represent TI inactivation of 75–80% (⊞), 80–85% (▨), 85–90% (◩), 90–95% (⊟), 95–100% (⊠) and above 100% (☐), respectively. The contour curves from left to right represent TI inactivation values of 80%, 85%, 90%, 95%, and 100%, respectively. It is obvious that the mass temperature has the most significant

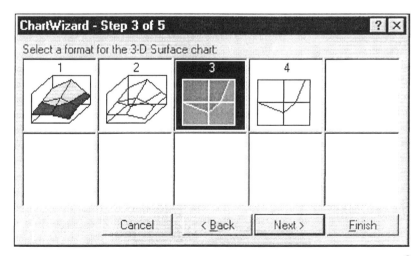

Figure 3.30 Select the suitable graphic type for generating a 2D contour plot using Ms-Excel®.

effect on TI inactivation, because all the contours are crossed if the temperature increases from 123°C to 177°C. The effect of temperature at low moisture is larger than that at high level, because at low moisture level (say at 16%) there are five crossings of contours on the temperature axis, whereas at high moisture level there are only four. This conclusion can also be achieved in this way: a line parallel to the moisture axis will cut across a contour line with higher TI inactivation at a lower moisture level and cut another contour line with lower TI inactivation at a higher moisture level.

The quasi-parallelism of these contour lines to the axis of the moisture content indicates that the moisture content affects TI inactivation not so significantly. Furthermore, it can be concluded from the size of the interval between two contour curves that the influences of the moisture content on the D_{TI} are more intensive at high than that at low mass temperature. These almost linear curves imply that no significant interaction effect exists between mass temperature and moisture content on TI inactivation.

3.7.1.4 3D + 2D Graphics

Some computer programs, such as STATISTICA and SYSTAT, provide the possibility to draw the 3D response plot as well as its projected contours on the bottom or on the top of the diagram in the same 3D diagram. Such diagrams can supply more information than a single 3D surface or 2D contour plot alone. To create such a plot a macro is usually required. Table 3.20 shows a macro for generating the graphic in Figure 3.31, which presents a response surface and the corresponding contours plotted on the bottom with SYSTAT. In addition, this macro can be easily modified to generate overlaid 2D contour plots or 3D contour surface graphics, which will be discussed in the following sections.

3.7.1.5 3D Contour Surface Plot

A graphic provides undoubtedly a great deal of directly understandable information about the relationship between the variables in a food system. Sometimes it is very helpful to include three independent variables in one graphic to analyze their effects on one food quality objective. The basic idea is to draw several contour surfaces within a 3D space, instead of representing only two independent variables and one response. The concept of doing this is relatively simple. It requires an algebraic transformation of the mathematical model. Instead of representing the quality index Z as a function of n independent factors $X_1, X_2, \ldots, X_i, \ldots, X_n$, one of the independent factors, say X_n, is represented as a function of Z, whereas the others remain variables $X_1, X_2, \ldots, X_i, \ldots, X_{n-1}$. Functional representation of the transformed relation is as follows:

$$X_n = f(Z, X_1, X_2, \ldots, X_{n-1}) \tag{3.24}$$

TABLE 3.20. Macro and Explanation for Generating of 3D + 2D Graphics Using SYSTAT.

Macro	Explanation
Sygraph	Start the graphic generating module
begin	
type=swiss	Set the character type
cs=1.8	Set the character size
use x3	Define the variables and original data
origin=5,3	Set the diagram relative position
Rem write title	Define the color for the title
color graph=black	Set the perspective of the title
depth=3IN	Select the surface on which the title be typed
facet=XZ	
write "n=140 min-1"/height=2.5 width=0.2IN x=1.5IN y=1.7IN center	Write the title at defined position with defined size
Rem draw the coordinate box	
plot z*y*x,	Draw the XYZ 3D coordinate box
/xmin=123,xmax=177, xlabel='Temperature (⁼C)', ymin=16,ymas=32, ylabel='Moisture (%)', zmin=70,zmax=100, zlabel='TI Inactivation (%)',	Define and write the axis labels
axes=12,size=0	Define the axis number in the coordinate
Rem draw the surface	Input the equation or model
plot Y3=47.23+0.00218*X1* X1+0.8923*X2-0.00858*X1*X2, !xmin=123,xmax=177, ymin=16, ymax=32,zmin=70,zmax=100, cut=20, xlabel='', ylabel='', zlabel='', axes=0, scale=0	Set the range of the three axis; cut defines the number of lines to be drawn within the surface
Rem draw the perspective contour curves	Define the projection surface and position on which the contours be drawn
facet=xy	
depth=0%	Select the character size
cs=2.0	
plot Y3=47.23+0.00218*x1*X1 +0.8923*x2-0.00858*x1*X2,	Intput the model for contour graphic
!contour,xmin=123,xmax=177, min=16,ymax=32,zmin=70, zmax=100, xlabel=' ', ylabel=' ', zlabel=' ', axes=0, scale=-5, ztick=5	Set the range of the three axes; select the number of contour curves
end	Finish the macro

99

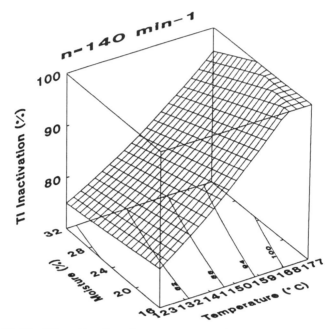

Figure 3.31 3D + 2D graphic: effect of mass temperature and mass moisture on TI inactivation.

Actually for the first-degree polynomial without [Equation (3.5)] and with interaction terms [Equation (3.6)] the transformed equations are, respectively, as follows:

$$X_n = Z - \beta_0 - \sum_{j=1}^{n-1}(\beta_j \cdot X_j) \tag{3.25}$$

$$X_n = \frac{Z - \beta_0 - \sum_{j=1}^{n-1}(\beta_j \cdot X_j) - \sum_{j<k=2}^{n-1}(\beta_{jk} \cdot X_j \cdot X_k)}{\beta_n + \sum_{j=1}^{n-1}(\beta_{jn} \cdot X_j)} \tag{3.26}$$

For a second-degree polynomial, there are two equations:

$$X_n = \frac{-B - \sqrt{B^2 - 4 \cdot A \cdot C}}{2 \cdot A} \tag{3.27}$$

$$X_n = \frac{-B + \sqrt{B^2 - 4 \cdot A \cdot C}}{2 \cdot A} \tag{3.28}$$

where

$$A = \beta_{nn}$$

$$B = \beta_n + \sum_{j=1}^{n-1}(\beta_{jn} \cdot X_j)$$

$$C = \beta_0 + \sum_{j=1}^{n-1}(\beta_j \cdot X_j) + \sum_{j=1}^{n-1}\left(\beta_{jj} \cdot X_j^2\right) + \sum_{j=1}^{n-2}\sum_{k=j+1}^{n-1}(\beta_{jk} \cdot X_j \cdot X_k) - Z$$

Given a specific value of the response Z, a response surface can be generated for X_n and any other two variables X_j and X_k according to Equations (3.25)–(3.28). For a given Z value, Equations (3.25) and (3.26) can be used to generate one surface at one time, whereas in a quadratic system two surfaces might be plotted according to Equations (3.27) and (3.28). These "response surfaces" are for the independent variable X_n, but not for the true Z. The Z value at one surface is constant, and this surface is really a contour surface. A number of contour surfaces can be generated in a 3D system by selecting different response values of interest. It is a good idea to select Z values evenly from its maximum and minimum in the actual experimental data.

If more than three factors are involved in the model, the original equation can still be used to create contour surface plots. The combination of X_n with two other variables X_j and X_k is selected for plotting, whereas the other variables are fixed at their constant levels (normally at their middle or extreme or any desirable levels).

A 3D contour surface supplies an overall behavior picture of the response in the entire experimental region. Such a diagram might supply more information than a normal 3D response surface, because three independent variables instead of two are involved. Of course, the information presented in 3D surface contours can also be given in 2D contours, but with less information about only two factors. A 3D contour surface is especially useful for a food system with only three independent variables. In this case, the optimum ranges—usually a maximum or minimum—can be visually found easily.

Unfortunately, few computer programs can be used to create a 3D contour surface chart directly. The macro described in Table 3.20 for SYSTAT can be easily modified to generate such graphics. One needs only to replace these commands instructing contour plotting with those of response surface drawing, which is also listed in the same macro, and change the corresponding equations. In general, 3D contour surface plots can be generated with a small trick: A normal computer program suitable to generate a response surface is used to create several "response surface" plots of one selected independent variable. These surfaces are then overlaid with help of a Draw or CAD program to generate the 3D surface contour plot. A practical example below shows this technique in detail.

As stated in Section 3.6.5, a model [Equation (3.22)] was built to study the inactivation of TI in soybeans. In the model, three important factors, namely, the mass temperature T, the mass moisture M, and the screw rotation speed n, are included. In a 3D contour surface the effects of these three factors on the TI inactivation D_{TI} can be described simultaneously. To generate such a plot, the first step is to rewrite Model 3.20 as follows:

$$n = (47.23 - D_{TI} + 0.00218 \times T^2 - 0.00858 \times T \times M)/(0.00638 \times M)$$
$$(3.29)$$

Then four suitable values of TI inactivation 70, 80, 90, and 100 are selected to create four additional equations [Equations (3.30)–(3.33)]. These values are between the minimal and maximal TI inactivation values of its actual data.
$D_{TI} = 70$:

$$n = (-22.77 + 0.00218 \times T^2 - 0.00858 \times T \times M)/(0.00638 \times M) \quad (3.30)$$

$D_{TI} = 80$:

$$n = (-32.77 + 0.00218 \times T^2 - 0.00858 \times T \times M)/(0.00638 \times M) \quad (3.31)$$

$D_{TI} = 90$:

$$n = (-42.77 + 0.00218 \times T^2 - 0.00858 \times T \times M)/(0.00638 \times M) \quad (3.32)$$

$D_{TI} = 100$:

$$n = (-52.77 + 0.00218 \times T^2 - 0.00858 \times T \times M)/(0.00638 \times M) \quad (3.33)$$

These four equations are now normal functions with only two independent variables T (123–177°C) and M (16–32%). The surface plot is made for each equation of them using the normal 3D response surface generating procedure within the same range of temperature and moisture content. Each surface is actually a diagram of one contour surface. These four surface plots are then copied or imported into Micrograx Designer® (or Charisma®, Corel Draw®, MS-PowerPoint®) and overlaid on each other. All coordinates but one are deleted to get a clean view of the contour surface graphic. While overlaying, one must be careful not to change the size of any surface plot; otherwise, the position of the contour surface may be incorrect in the final diagram.

The 3D contour surface plot resulting from overlaying is illustrated in Figure 3.32. It is clear to see that a total TI inactivation can be achieved in the room indicated, namely, at high mass temperature and screw rotation speed but low mass moisture content. By contrast, at low temperature and low screw

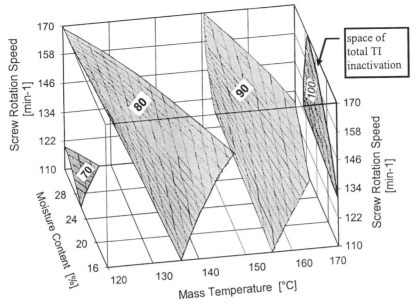

Figure 3.32 A 3D contour surface plot: effect of mass temperature mass moisture and screw speed on TI inactivation.

rotation speed but with high moisture content the TI will be protected from inactivation.

3.7.1.6 Software

A series of popular computer programs can be applied to generate each type of graphics mentioned above. Some of them and their corresponding features are listed alphabetically in Table 3.21, and their full communication addresses can be found at the end of Chapter 1.

3.7.2 NUMERICAL METHOD

The effects of independent variables on food quality can also be analyzed numerically. The essential technical background of the numerical method is actually similar to that of the graphical approach. In this method the response values are calculated without being further plotted. These computed data can be individually viewed and analyzed to understand the effects of each variable on the response. Compared with the graphical method, the numerical method cannot provide the food researcher with an overall visual survey of all the relationships between variables and responses.

TABLE 3.21. Computer Programs Supporting Graphic Generating.

Software	X-Y Plot	2D Contour	3D Surface	3D + 2D Plot	3D Contour
APO®	+	+			
Corel Chart®	+	+	+		
Harvard Graphic®	+	+	+		
JMP®	+	+	+		
Lotus 1-2-3®	+	+	+		
Lotus Improve®	+	+	+		
Micrograx Charisma®	+	+	+		
Minitab®	+	+	+		
Ms-Excel®	+	+	+		
MODDE®	+	+	+		
SAS®	+	+	+		+
SPSS®	+	+	+		+
SigmaPlot®	+	+	+	+	
Standford Graphics®	+	+	+	+	
StatGraphics®	+	+	+		
STATISTICA®	+	+	+	+	+
SYSTAT®	+	+	+	+	+
TableCurve	+	+	+	+	

*The symbol "+" means a program has this feature.

3.8 PREDICTION AND OPTIMIZATION

Using the mathematical model to predict the food quality under different level combinations of independent variables is very attractive. It enables the food researchers to reveal the relationship between food quality and independent variables in the regions where actually no trials have been performed. In food product development the term prediction has two meanings:

- to calculate the response values according to certain levels of independent variables
- to calculate the possible independent variable levels with a given response value

The prediction is purely a mathematical calculation based on the model or equation built. Usually a small computer program in BASIC is helpful to perform these prediction calculations.

A further or specific step of prediction is optimization, which is possibly the most important step and one of the final targets of the statistical food product development. Recently, the term optimization has become increasingly popular in many sectors of the consumer products industry. It appears to have similar meanings, depending on the perspectives in food product development, food quality management, marketing, and so on. However, in food product design the optimization is intended to provide a more precise map of the path to the highest probability for a successful food production, to improve the overall effectiveness of research and development, and to enhance its role in the food product design. During the analysis of effects of the important variables, the contributions of variables to the food quality indices have been identified. However, optimization supplies more detailed information about the level combinations of independent variables that will yield optimum food quality. Of course, this information from optimization is given only if the model built is significant and accurate. It is usual that there are several sets of variable levels (combinations) to produce foods with most desirable quality attributes. In other words, an optimum product may be achieved with different combinations of the variable levels. The graphical approaches and numerical methods can be used to obtain the optimum and will be introduced in the following sections.

When doing prediction and optimization, it must be kept in mind that the prediction and optimization results are correct with a certain statistical probability. There can never be a prediction or optimization made with absolute certainty; therefore, the reliability and the correctness of the prediction or optimization results should always be tested and checked through practical validation and confirmations.

3.8.1 GRAPHICAL APPROACH

The graphical method for optimization reduces drastically the amount of time and effort required for investigation of multiresponse or multifactor systems. This method provides a comprehensive and informative insight into the system and leads to fast process optimizations. Furthermore, it is easily adaptable to most commonly used computer software packages, and the diagrams created may be used as guidelines during food processing, quality control, and product development and improvement.

3.8.1.1 System with Two Variables

In the graphical approach, predictive models are used to create contour lines within the experimental region on two-dimensional diagrams. These plots present information for two factors and one or more responses and are reasonably accurate within the experimental region. Of course, the practical plot accuracy also depends on the representational accuracy (p or R^2) of the model. The regions of optimum response(s) are judged by visual inspection of the contour lines. This method reduces the possibilities of "unrealistic" solutions, because only the region within the experimental space is examined, and it allows simultaneous optimizations of several competing responses by simple superimposition.

Figure 3.33 shows a 2D contour plot as an illustration. From the diagram it can be concluded that the maximal response value of more than 4.9 would

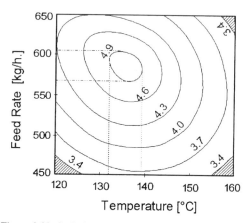

Figure 3.33 Optimization with the 2D graphical approach.

be achieved in the room of "132°C < temperature < 138°C, 570 kg/h < feed rate < 610 kg/h," but the minimal responses locate in the three corners outside the contour lines ▨ with the response value of 3.4 and have the response value smaller than 3.4.

3.8.1.2 Common Optimum

Usually, a food product has more than one important quality index, which must be considered simultaneously at the stage of optimization. In this case, to optimize all the quality aspects of the food to find the common optimum region is the goal of product developers. Normally, these food quality indices (responses) are not equally important. Optimization in such a food system is actually a compromise of all these quality properties, because there is seldom a combination of variable levels that yields the best value for all the food quality indices. To do optimization in this case, the more important response(s) is considered first and given a high weight factor, so that its best or nearly its best value may be achieved according to the optimization variable levels.

The 2D contour plot has proved convenient to solve optimization problems with common optima. The contour lines of the different quality indices based on their corresponding models are plotted and overlaid into a single diagram. The optimum region(s) can then be examined and chosen visually. Figure 3.34 illustrates such an overlaid contour plot of two responses Z1 and Z2. If the common optimum requires a maximum of both Z1 and Z2, the region A is the suitable optimum region, with Z1 > 4.9 and 22 > Z2 > 20. If the common optimum should be a minimum of Z1 and Z2, then the region B indicates the unique optimum area (Z1 < 3.4 and Z2 < 2). Similarly, if the optimum requires a Z2 value smaller than 16 but a maximal Z1, then the region C is the correct optimum region.

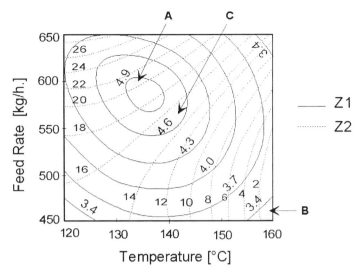

Figure 3.34 Graphical optimization to find common optima.

3.8.1.3 System with Three Variables

A 3D contour surface plot allows the simultaneous description of three factors in one chart. It can be easily used to make a graphical optimization of a food system with three factors. A number of 3D contour surfaces are generated in a 3D space according to the principles described in Section 3.7.1.5, and the optimum region can be found and checked visually. For example, in Figure 3.32 the region of a total inactivation of the TI is located at the front-upper-right corner in the diagram, namely, in the range of high temperature and high screw rotation speed but low moisture content. However, this 3D graphical approach is usually not suitable for optimization with common optima.

3.8.1.4 Optimization for System with More Than Three Variables

For a food system with four or more factors, the graphical approaches are in principle no longer suitable to do optimization, unless they are modified to illustrate the optimum regions. In a 3D contour surface plot only three factors can be involved in optimization. The fourth or any further variable must be set at constant values that are actually their optimum levels. The optimum levels of these variables that are not included in the 3D contour surface plot must be figured out first. These variables in the full model are then replaced with their corresponding optimum levels to obtain a simplified model with only three factors. This model can then be used to do optimization through the 3D contour surface plotting approach as described above. If a 2D contour line diagram is used in optimization, then all the factors but two that are selected must be set at their optimum levels following the similar principles as discussed above.

3.8.2 NUMERICAL METHOD

The numerical or calculation method is a most universal optimization approach. Although it supplies no obvious overall information about the whole system, it can perform complicated optimizations such as to optimize a system with more than two or three independent variables and several responses. The optimum level of interesting combinations of independent variables could be directly determied with the help of a small computer program. Usually, such a program can be written by the food researcher himself in BASIC, FORTHEN, PASCAL, C or any other languages. Among them, BASIC and its variants are quite simple and widely used programming languages. It is generally supplied together with Ms-DOS 3.X–6.X and available almost everywhere. Because of the high performance of a modern computer, the calculations can usually be completed within several seconds, no matter how complicated the optimization may be and which programming language is used.

The easiest way in seeking optimum solutions in a food system is to use the scanning principle. The response values are calculated at all possible level combinations of the independent variables, which vary from their smallest to their largest values. These response values are compared with each other to decide which level combinations of the independent variables lead to the desired response values. These combinations are, in principle, the optimum. Table 3.22 shows a small program in BASIC which can be used to seek the maximal optimum in a single response system with four variables X_1, X_2, X_3, and X_4.

Classical theory on maximum and minimum can also be used in the optimization if the optimum is a maximum or a minimum. There are also numerous optimization methods developed by various mathematicians. The scanning method is, however, most widely used because of its easy programming and simple structure. On a modern computer, the performance differences between the scanning method and any others may be just a matter of a few seconds.

3.9 EXPLORATION AND CONFIRMATION

Statistical food product design is often an interactive procedure that entails going back and forth between the real world of experimentation and the theoretical world of modeling and optimization. Each step gives direction to the next, until desired optimum products are achieved. This means that the optimum combinations of variables must be checked and confirmed in practice. The properties of the optimized food products must be tested to validate the correctness and accuracy of the model as well as of the optimization.

TABLE 3.22. **A BASIC Optimization Program.**

```
REM A small program for optimization - scanning for maximum
REM X1 varies from X1_min to X1_max; X2 varies from X2_min to X2_max
REM X3 varies from X3_min to X3_max and X4 varies from X4_min
  to X4_ max
cls ' clear the screen

REM Define the file "optimum 1.dat" for output of optimum results
OPEN "Optimum 1.dat" FOR OUTPUT AS #1
PRINT #1, " X1 "; " X2 "; " X3 "; " X4 "; " Z "
REM Set the step of scanning at 20. It can be bigger than 20, if more accurate
REM optimum is hoped, otherwise it can be smaller than 20
X1_step = (X1_max - X1_min)/20
X2_step = (X2_max - X2_min)/20
X3_step = (X3_max - X3_min)/20
X4_step = (X4_max - X4_min)/20

REM Calculate the starting value of response Z0
Z0 = f(X1_min, X2_min, X3_min, X4_min)

REM start the scanning procedure
for X1 = X1_min to X1_max step X1_step
  for X2 = X2_min to X2_max step X2_step
    for X3 = X3_min to X3_max step X3_step
      for X4 = X4_min to X4_max step X4_step

REM calculate the response value at each combination of variables
Z = f(X1, X2, X3, X4)

IF Z < Z0 THEN GOTO 100
PRINT #1, X1;",";X2; ",";X3; ",";X4, " ";Z
100        NEXT X4
         NEXT X3
       NEXT X2
     NEXT X1
     BEEP
     CLOSE #1
End
```

Extra trials other than the initial designed trials included in the experimental plan are known as checkpoints. They are usually designed additionally and performed to check the reliability and validity of the model. They are normally symmetrically selected combinations of the midlevels between two tested levels for each variable in the original experimental design. The real response values at the checkpoints will be compared with the values predicted by the model at the same checkpoints to decide whether the model is exact and practically usable. If the differences between the real and the predicted response values are fairly small and acceptable, then the model is assumed adequate for prediction and optimization purposes. Otherwise, the model must be improved.

In practice, the trials of interest, usually the optimum combination of the variables, are also used to examine the practical reliability of the model. If the real product obtained appears not to be the optimum, then questions can be asked to find the causes. A troubleshooting guide is listed in Figure 3.35, which may be used to determine the causes for the inacurate prediction.

3.10 TRANSFER OF OPTIMUM RESULTS AND EVOLUTIONARY OPERATION

Normally, a food product design is performed in a laboratory and in a pilot plant. The final objective of food product design is, however, to produce the specific food under industrial conditions. In other words, the optimum result from the laboratory or pilot plant equipment must be further transferred to a full-scale industrial production process. It is well known that a large performance difference exists between the equipment in the laboratory versus that in industrial production. For example, the mixing effect of an industrial mixer with a large volume is lower, the flow rate and the flow behavior of liquid materials are changed, and the speed of heating up and cooling down in an industrial reactor may be delayed. These kinds of differences may cause divergence or distortion between the optimums of a small unit compared with an industrial process. The optimum obtained in the laboratory cannot simply be regarded as the optimum for an industrial process of a similar type. Even if the full-scale plant begins operation at the optimum, it can eventually "drift" away from that point because of variations in raw materials, environmental changes, and different operating behavior of the personnel.

To solve the problems that occur during transfer of optimum results obtained from small-scale equipment to an industrial scale process, a statistical technique of the so-called Evolutionary Operation (EVOP), proposed originally by Box in 1957, proved to be helpful. The basic thought of Box during his development of this method was to run a process in such a way to generate product plus information on how to improve the product. In fact, there is normally no fundamental difference between a small and a large scale process of the same type with respect to the qualitative relationship between the food quality responses and the independent variables. What may frequently occur is the drifting of the optimum region of the food quality response as shown in Figure 3.36.

From a practical point of view, EVOP consists of systematically introducing slight changes to the levels of the independent variables that are large enough to induce detectable response differences but not large enough to generate any off-specification productions. While applying EVOP, 2^n or 2^{n-p} factorial experimental plans are normally designed in advance for the industrial process, with

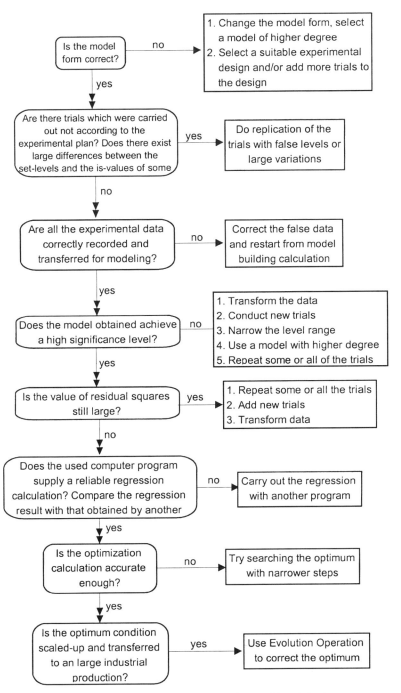

Figure 3.35 Troubleshooting guide to confirm model optimization.

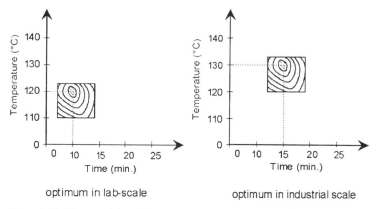

optimum in lab-scale   optimum in industrial scale

Figure 3.36 Changes of the optimum region from a lab to an industrial scale process.

the test levels of each variable being closely and symmetrically located around the original optimum condition obtained from laboratory or pilot equipment. The span of test levels must be small enough so that a level variation would not cause any noticeable change in the food quality or interruption of production, which will still permit the estimate of the standard error. When one observation has been taken at each design point, a cycle is completed. The effects and interactions of the process variables may then be computed. Eventually, after several cycles, the effect of one or more process variables, or their interactions, may appear to have a significant effect on the response that indicates the optimum drift direction as well. At this point, a decision may be made to change the basic operations to improve the response. When improved conditions have been detected, a phase is completed. A new factorial design is then set up at the new "optimum" in the same way to check and detect the new optimum drift direction of further variable(s). Repeat this action until no significant variation can be discovered. The last optimum will be the real optimum for this industrial production process.

In testing the significance of process variables and interactions, an estimate of experimental error is required. This is calculated from the cycle data. The 2^n design is usually centered about the current best operating conditions, so, if the process is really centered at the maximum, the response at the central point should be significantly greater than those responses at the 2^n-peripheral points.

There are other optimization techniques such as the One-Factor-at-a-Time Method, the Simplex-Search Method, and the Steepest Ascent Method, that can be used to find the new or drifted optimum region. The EVOP is the simplest way to gain information for the correction of the optimum region, and can be performed according to a standardized worksheet. Theoretically, the EVOP can

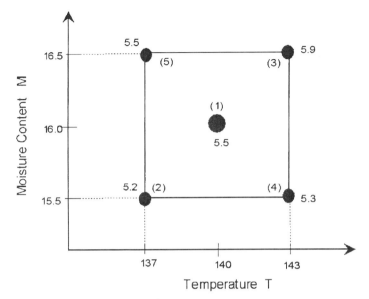

Figure 3.37 2^2 factorial design for EVOP.

be applied to n process variables. For simplicity, however, it is performed with only two to three variables. The EVOP should be frequently performed at an industrial production line so that the real optimum can always be followed, no matter what variation may occur in production. The following is an example of the EVOP procedure for two variables.

Consider the food cooking extrusion whose expansion ratio of the extrudates is a function of mass temperature (T) and moisture content (M). The current optimal operating conditions are $T = 140°C$ and $M = 16\%$ with an extrudate expansion ratio of 5.5. The EVOP procedure uses the 2^2 factorial design plus center point shown in Figure 3.37. The cycle is completed by running each design point in numerical order [(1), (2), (3), (4), and (5)]. The expansion ratio in the first cycle are also shown in Figure 3.37 and are entered in the EVOP calculation sheet, shown in Table 3.23. At the end of the first cycle, no estimate of the standard deviation can be made. The effects and interaction of the temperature and moisture content are calculated in the usual manner for a 2^2 factorial design.

A second cycle is then run, and the expansion ratio data are entered in another EVOP calculation sheet (Table 3.24). At the end of this cycle, the experimental error can be estimated. Note that the "Range of (iv)" refers to the range of the maximal differnces in row (iv); thus the Range of (iv) is $1.0 - (-1.0) = 2.0$. Because none of the effects in Table 3.24 exceeds their error limits, the true effect

TABLE 3.23. EVOP Calculation Sheet and Data in the First Cycle ($m = 1$).

5 • →T 3 M •1 2• •4	Cycle: $m = 1$ Response: Expansion Ratio		Phase: 1 Date: 12.03.95 Page: 1

	Calculation of Averages		
Operating Conditions	(1) (2) (3) (4) (5)		Standard Deviation
(i) Previous cycle sum	5.5 5.2 5.9 5.5 5.3		Previous sum $S =$
(ii) Previous cycle average			Previous average $S =$
(iii) New observations			New $S =$ Range · $f_{5,m}$ $=$
(iv) Differences [(ii) − (iii)]			Range of (iv) $=$
(v) New sums [(i) + (iii)]	5.5 5.2 5.9 5.5 5.3		New Sum S $=$
(vi) New averages [$y_i = (v)/n$]	5.5 5.2 5.9 5.5 5.3		New average S $=$ new sum $S/(n − 1)$
	Calculation of Effects		Calculation of Error Limits
Temperature effect $= 1/2(\bar{Y}_3 + \bar{Y}_4 − \bar{Y}_2 − \bar{Y}_5)$ Moisture effect $= 1/2(\bar{Y}_3 + \bar{Y}_5 − \bar{Y}_2 − \bar{Y}_4)$ $T \cdot M$ interaction effect $= 1/2(\bar{Y}_2 + \bar{Y}_3 − \bar{Y}_4 − \bar{Y}_5)$ Change-in-mean effect $= 1/5(\bar{Y}_2 + \bar{Y}_3 + \bar{Y}_4 + \bar{Y}_5 − 4 \cdot \bar{Y}_1)$			For new average $2S/\sqrt{n}$ For new effects $2S/\sqrt{n}$ For change in mean $1.78S/\sqrt{n} =$

is probably zero, and no changes in operating conditions are contemplated. The table values of $f_{k.m}$ are listed in the appendix, where k is the trial number that usually equals $(2^n + 1)$.

The results of a third cycle are shown in Table 3.25. The effect of moisture content (0.87) now exceeds its error limit, and the temperature effect is equal to the error limit (0.67). A change in operating conditions is now probably justified. In light of the results, it seems reasonable to begin a new EVOP phase around point (3). Thus, $T = 143°C$, $M = 15.5\%$ is the improved optimum and would become the center of 2^2 factorial design in the second phase of EVOP.

3.11 A COMPLETE PRACTICAL EXAMPLE

To show the entire perspective of statistical food product design, a completed example is given as follows. This example was a research project undertaken to tackle problems associated with shrimp cracker development and optimization through cooking extrusion.

TABLE 3.24. EVOP Calculation Sheet and Data in the Second Cycle ($m = 2$).

	Cycle: $m = 2$					Phase: 1
	Response: Expansion Ratio					Date: 12.03.95
						Page: 2

Calculation of Averages						
Operating Conditions	(1)	(2)	(3)	(4)	(5)	Standard Deviation
(i) Previous cycle sum	5.5	5.2	5.9	5.5	5.3	Previous sum $S =$
(ii) Previous cycle average	5.5	5.2	5.9	5.5	5.3	Previous average $S =$
(iii) New observations	5.9	5.6	6.9	4.5	5.0	New $S =$ Range $\cdot f_{5,m}$ $= 2.0 \cdot 0.30 = 0.60$
(iv) Differences [(ii) − (iii)]	−0.4	−0.4	−1.0	1.0	0.3	Range of (iv) $= 1.0 - (-1.0) = 2.0$
(v) New sums [(i) + (iii)]	11.4	10.8	12.8	10.0	10.3	New sum S $= 0.60$
(vi) New averages [$yi = (v)/n$]	5.7	5.4	6.4	5.0	5.15	New average $S =$ new sum $S/(n-1) = 0.60$

Calculation of Effects		Calculation of Error Limits
Temperature effect $= 1/2(\bar{Y}_3 + \bar{Y}_4 - \bar{Y}_2 - \bar{Y}_5)$	0.43	For new average $2S/\sqrt{n}$ $= 0.85$
Moisture effect $= 1/2(\bar{Y}_3 + \bar{Y}_5 - \bar{Y}_2 - \bar{Y}_4)$	0.58	For new effects $2S/\sqrt{n}$ $= 0.85$
$T \cdot M$ interaction effect $= 1/2(\bar{Y}_2 + \bar{Y}_3 - \bar{Y}_4 - \bar{Y}_5)$	0.83	
Change-in-mean effect $= 1/5(\bar{Y}_2 + \bar{Y}_3 + \bar{Y}_4 + \bar{Y}_5 - 4 \cdot \bar{Y}_1)$	−0.17	For change in mean $1.78 S/\sqrt{n} = 0.76$

Shrimp crackers, which are conventionally produced by manual labor, are one of the most famous snack foods in the Far East. Food cooking extrusion is a modern compact food-processing technology with versatile advantages. This study tried to produce snack foods similar to shrimp crackers through cooking extrusion. The original recipe for commercial manufacture of shrimp crackers was used here without any modification. A Clextral twin-screw corotating extruder (Type BC 45 with three zones) was adapted to produce the snacks and the following four independent variables were controlled:

- mass temperature T (°C)
- screw rotation speed n (min^{-1})
- feed rate V_f (g d.s. min^{-1})
- mass moisture content M (%)

The experiments were designed according to the principle of CCD, including the variables in Table 3.26. Five replications at the central point were designed

TABLE 3.25. EVOP Calculation Sheet and Data in Third Cycle ($m = 3$).

(diagram)	Cycle: $m = 3$ Response: Expansion Ratio					Phase: 1 Date: 12.03.95 Page: 3

	Calculation of Averages					
Operating Conditions	(1)	(2)	(3)	(4)	(5)	Standard Deviation
(i) Previous cycle sum	11.4	10.8	12.8	10.0	10.3	Previous sum $S = 0.60$
(ii) Previous cycle average	5.7	5.4	6.4	5.0	5.15	Previous average $S = 0.60$
(iii) New observations	6.0	5.0	7.6	5.9	6.2	New S = Range · $f_{5,m}$ $= 1.60 \cdot 0.35 = 0.56$
(iv) Differences [(ii) − (iii)]	−0.3	0.4	−1.2	−0.9	1.05	Range of (iv) $= 0.4 - (-1.2) = 1.60$
(v) New sums [(i) + (iii)]	17.4	15.8	20.4	15.9	16.5	New sum $S =$ $= 0.60 + 0.56 = 1.16$
(vi) New averages [$yi = (v)/n$]	5.8	5.27	6.8	5.3	5.5	New average $S =$ new sum $S/(n-1) = 0.58$

Calculation of Effects		Calculation of Error Limits
Temperature effect $= 1/2(\bar{Y}_3 + \bar{Y}_4 - \bar{Y}_2 - \bar{Y}_5)$	0.67	For new average $2S/\sqrt{n}$ $= 0.67$
Moisture effect $= 1/2(\bar{Y}_3 + \bar{Y}_5 - \bar{Y}_2 - \bar{Y}_4)$	0.87*	For new effects $2S/\sqrt{n}$ $= 0.67$
$T \cdot M$ interaction effect $= 1/2(\bar{Y}_2 + \bar{Y}_3 - \bar{Y}_4 - \bar{Y}_5)$	0.64	
Change-in-mean effect $= 1/5(\bar{Y}_2 + \bar{Y}_3 + \bar{Y}_4 + \bar{Y}_5 - 4 \cdot \bar{Y}_1)$	−0.07	For change in mean $1.78S/\sqrt{n} = 0.60$

to examine the experimental error. The α value was calcualted according to Equation (3.13) and was 1.664 (see Figure 3.9). The real and coded test levels of each variable were chosen according to results of some preliminary trials and on the basis of experience (Table 3.26). The real levels of each variable T, n, V_f, and M are transformed into coded test levels ($+1, -1, 0, +\alpha, -\alpha$)

TABLE 3.26. Parameters of Extrusion—Basis for the Experimental Plan.

Variable	T (°C)	n (min⁻¹)	V_f (g/min)	M (%)
Coded variable	X_1	X_2	X_3	X_4
Middle level (0)	75	60	223	40
Upper level (+1)	90	80	295	50
Lower level (−1)	60	40	150	30
Upper star point (+α)	100	93	343	56.5
Lower star point (−α)	50	27	102	23.5

by Equations (3.34)–(3.37). The four factors are accordingly changed into the coded variables X_1, X_2, X_3, and X_4, respectively.

$$X_1 = \frac{(T - 75) \cdot 2}{90 - 60} \tag{3.34}$$

$$X_2 = \frac{(n - 60) \cdot 2}{80 - 40} \tag{3.35}$$

$$X_3 = \frac{(V_f - 223) \cdot 2}{295 - 150} \tag{3.36}$$

$$X_4 = \frac{(M - 40) \cdot 2}{50 - 30} \tag{3.37}$$

All the single trials in the experimental plan were performed according to the buildup principle of the CCD. First, the basic 2^4 factorial design (no. 1–16 in Table 3.27) together with the five repeated trials at the central point (no. 17–21) would be realized. A corresponding mathematical model involving the four factors and their interaction effects would then be designed. If the model built in this way were unable to reach a desired significance level through the F-test, then those trials at the star points (no. 22–29) must be performed so that a second-degree model can be further built.

Based on observed preferences in shrimp cracker consumption in Asia, the specific volume D_{sp} (ml/g) and the maximal shear force F_{br} (N) of the fried extrudates were selected as the objective snack quality indices. Therefore, the specific volume is expected to be as large as possible, whereas the maximal shear force should be similar to that of a commerical shrimp cracker ($F_{br0} = 5.03$), which was awarded a gold medal in 1988 by the Chinese National Committee for Products in the Light Industry.

A couple of flat dies with a dimension of $0.65 \times 20.5(H \times L)$ mm^2 were used for the cooking extrusion. During experimentation, the extrudates were cut into pieces of about 4 cm in length and further dried to a moisture content of about 9.5% in a ventilated oven at 50°C. The dry extrudates were fried in an oil bath at 180°C for approximately 40 seconds until they were thoroughly expanded. For the determination of the specific volume D_{sp}, some 25–40 g fried samples were used according to the volume displacement method used for bread volume measurement. A universal rheometer (Type NRM-2002 J, Fudoh Kogyo, Japan) was applied to examine the maximal shear force of the fried extrudates. For each single trial, 30 samples were used to determine their maximal shear force. The average value of their sum, but without the largest and the smallest ones, was taken as the real maximal shear force, F_{br}. The two-response measurement results are also presented in Table 3.27.

TABLE 3.27. Real and Coded Experimental Plan and the Data of D_{sp} and F_{br}.

No.	Coded Experimental Plan				Real Experimental Plan				Responses	
	X_1	X_2	X_3	X_4	T	N	V_f	M	D_{sp}	F_{br}
1	1	1	1	1	90	80	295	50	3.62	11.73
2	1	1	1	−1	90	80	295	30	2.54	20.91
3	1	1	−1	1	90	80	150	50	7.99	5.08
4	1	1	−1	−1	90	80	150	30	3.69	12.55
5	1	−1	1	1	90	40	295	50	2.82	8.12
6	1	−1	1	−1	90	40	295	30	2.48	15.75
7	1	−1	−1	1	90	40	150	50	8.12	7.13
8	1	−1	−1	−1	90	40	150	30	4.15	11.76
9	−1	1	1	1	60	80	295	50	0.72	3.06
10	−1	1	1	−1	60	80	295	30	1.27	13.86
11	−1	1	−1	1	60	80	150	50	1.18	4.13
12	−1	1	−1	−1	60	80	150	30	2.62	12.15
13	−1	−1	1	1	60	40	295	50	0.71	3.67
14	−1	−1	1	−1	60	40	295	30	1.48	9.35
15	−1	−1	−1	1	60	40	150	50	1.38	4.95
16	−1	−1	−1	−1	60	40	150	30	3.31	16.75
17	0	0	0	0	75	60	223	40	3.05	10.06
18	0	0	0	0	75	60	223	40	3.04	10.03
19	0	0	0	0	75	60	223	40	2.95	10.01
20	0	0	0	0	75	60	223	40	2.98	9.98
21	0	0	0	0	75	60	223	40	3.02	10.02
22	+α	0	0	0	100	60	223	40		
23	−α	0	0	0	50	60	223	40		
24	0	+α	0	0	75	93	223	40		
25	0	−α	0	0	75	27	223	40		
26	0	0	+α	0	75	60	343	40		
27	0	0	−α	0	75	60	102	40		
28	0	0	0	+α	75	60	223	56.5		
29	0	0	0	−α	75	60	223	23.5		

A mathematical model for the specific volume, D_{sp}, was built through regression based on the D_{sp} results and the coded experimental plan in Table 3.27 [Equation (3.38)]. Calculation was performed with the help of the program SPSS/PC+. It describes the relationship between the specific volume of the fried extrudates and the four coded independent variables. The backward regression procedure was adopted, and this part of the output including the selected simplified model is listed in Table 3.28. As shown in the table, this model achieves an extra high significance level of zero through the F-test and a R^2 value of approximately 0.93. This means that this model is workable and can be applied in the subsequent prediction and optimization stages. It was, thus, not necessary to perform the additional trials at the star points (trial no. 22–29), which would support building a quadratic

TABLE 3.28. A Part of the Backward Regression Results for D_{sp}.

```
-------------------------------------------------------------------------------------------
Page 12                              SPSS/PC+                                      5/2/96
                         **** MULTIPLE REGRESSION ****
Equation Number 1        Dependent Variable: Y

Variable(s) Removed on Step Number
13:   X2X4

Multiple R               .96378
R Square                 .92887
Adjusted R Square        .89057
Standard Error           .64377

Analysis of Variance
                    DF           Sum of Squares        Mean Square
Regression           7              70.35400            10.05057
Residual            13               5.38771              .41444

F = 24.25099                    Signif F = .0000
-------------------------------------------------------------------------------------------
Page 13                              SPSS/PC+                                      5/2/96
                         **** MULTIPLE REGRESSION ****
Equation Number 1        Dependent Variable: Y
--------------------------Variables in the Equation----------------------
```

Variable	B	SE B	Beta	T	Sig T
X3X4	-.30000	.16094	-.13788	-1.864	.0851
X2X3	.13375	.16094	.06147	.831	.4210
X1X4	.89875	.16094	.41308	5.584	.0001
X1X3	-.51125	.16094	-.23498	-3.177	.0073
X4	.31250	.16094	.14363	1.942	.0742
X3	-1.05000	.16094	-.48259	-6.524	.0000
X1	1.42125	.16094	.65323	8.831	.0000
(Constant)	3.00571	.14048		21.396	.0000

```
-------------------------------------------------------------------------------------------
Page 14                              SPSS/PC+                                      5/2/96
```

model.

$$D_{sp} = 3.01 + 1.42 \times X_1 - 1.05 \times X_3 + 0.31 \times X_4 - 0.51 \times X_1 \times X_3$$

$$+ 0.90 \times X_1 \times X_4 + 0.13 \times X_2 \times X_3 - 0.3 \times X_3 \times X_4 \qquad (3.38)$$

From the size of each coefficient in Equation (3.38), it can be determined which term contributes most to the specific volume. In this case, it is the mass temperature (corresponding to X_1) which is very important for the specific volume, whereas the screw rotation speed (corresponding to X_2) plays a rather irrelevant role. In the 3D contour surface diagram, Figure 3.38, the effects of the three highly significant factors, T, V_f, and M, are illustrated according to

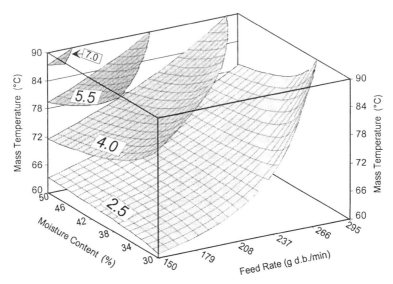

Figure 3.38 3D contour surface plot of D_{sp}: effects of mass temperature T, feed rate V_f, and moisture content M on the maximal specific volume D_{sp}.

Equation (3.38). The screw rotation speed n was held at $60\,\text{min}^{-1}$, that was, $X_2 = 0$. As can be seen, D_{sp} increases quickly with increasing mass temperature and increases more gradually with increasing moisture content but decreases with increasing feed rate. Thereafter, the snack with a maximal specific volume would be extruded at high temperature and with high moisture content but at a low feed rate.

In a similar way a mathematical model was set up for the maximal shear force, F_{br} (Equation 3.39). The related part of the backward regression results, in which the properties and the coefficients of the model were included, is shown in Table 3.29. This model achieves a high significance level higher than 0.001 by the F-test and a high R^2 value of about 0.95. Therefore, additional experimentation with trials 22–29 is no longer necessary.

From the size of each coefficient in Equation (3.39), it is apparent that mass temperature, feed rate, and moisture content are the most important factors for the maximal shear force. In a similar way as Figure 3.38 is plotted, their effects are illustrated in Figure 3.39 in a 3D contour surface diagram with $X_2 = 0$. The feed rate is obviously the most important factor influencing the maximal shear force, F_{br}, followed by the mass temperature and the moisture content. Higher mass temperature and moisture content at a low feed rate would lead to extrudates with larger maximal shear force, that is, a hard snack product. Notice that the two contour surfaces with an F_{br} value of 5 and 9 go in opposite directions. This means that between these two surfaces there must be a large

TABLE 3.29. A Part of the Backward Regression Results for F_{br}.

```
-----------------------------------------------------------------------
           **** MULTIPLE REGRESSION ****
Equation Number 1       Dependent Variable: Y
Variable(s) Removed on Step Number
14: X1X4
Multiple R              .97234
R Square                .94544
Adjusted R Square       .92205
Standard Error          1.27074

Analysis of Variance
                   DF     Sum of Squares    Mean Square
Regression          6       391.70644        65.28441
Residual           14        22.60676         1.61477
F = 40.42957             Signif F = .0000
-----------------------------------------------------------------------
```

| Page 16 | | SPSS/PC+ | | 5/2/96 | |

```
           **** MULTIPLE REGRESSION ****
Equation Number 1       Dependent Variable.. Y
-------------------------Variables in the Equation----------------------
```

Variable	B	SE B	Beta	T Sig	T
X2X3	1.20938	.31768	.23766	3.807	.0019
X1X3	1.75187	.31768	.34427	5.515	.0001
X1X2	.56437	.31768	.11091	1.777	.0974
X4	-4.07563	.31768	-.80092	-12.829	.0000
X3	.74687	.31768	.14677	2.351	.0339
X1	1.56937	.31768	.30841	4.940	.0002
(Constant)	10.05000	.27730		36.243	.0000

difference in the variable effects.

$$F_{br} = 10.25 + 1.57 \times X_1 + 0.75 \times X_3 - 4.08 \times X_4 + 0.56$$
$$\times X_1 \times X_2 + 1.75 \times X_1 \times X_3 + 1.21 \times X_2 \times X_3 \quad (3.39)$$

Snack food development is the final objective of this work. Hence, the two significant models should be used to optimize the extrusion process, namely, to find those variable levels that would lead to favorable extrudate quality, with maximal specific volume and maximal shear force of about 5.03 N. Thus, the optimization calculation Equation (3.39) was kept as a limitation and Equation (3.38) was used in scanning the maximal D_{sp} value and the corresponding coded levels of these four factors. A small computer program (Table 3.30) is written in BASIC for doing the optimization. The results of this program are written in the file "optimum 1.dat" as shown in Table 3.31.

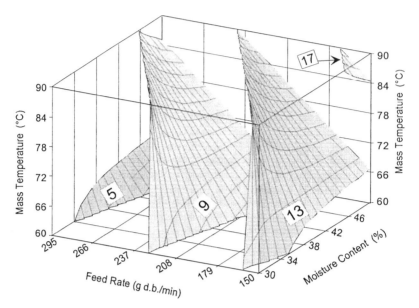

Figure 3.39 3D contour surface plot of D_{sp}: effects of mass temperature T, feed rate V_f, and moisture content M on the maximal shear force F_{br}.

The optimization result of running the program in Table 3.30 is listed in Table 3.31. Line no. 18 in Table 3.31 gives the final optimization result desired, namely, a fried extrudate with quality criteria of D_{sp} of approximately 7.48 (ml/g), and F_{br0} of 5.03 (N). It can be extruded by using the coded extrusion condition of $X_1, X_2, X_3, X_4 = (1, 0.05, -1, 0.99)$. This optimum region is also illustrated in the 2D contour diagram in Figure 3.40 by holding X_2 and X_4 at their optimum levels of 0.05 and 0.994, respectively. Through conversion of these coded levels according to Equations (3.34–3.37), the real optimum variable levels are calculated and listed below:

$$T_{optm} = 90(°C) \qquad V_{f.optm} = 150(g/min)$$
$$n_{optm} = 61(rpm) \qquad M_{optm} = 50(\%)$$

Extrusion under these conditions was performed to confirm the correctness and accuracy of the modeling and optimization. The quality of the extruded snacks was determined and is shown below. These quality indices of the extruded snacks indicate that a statistical approach led to successful food product development.

$$D_{optm.Extr.} = 7.5(ml/g)$$
$$F_{optm.Extr.} = 5.21(N)$$

TABLE 3.30. A Small BASIC Program for Snack Quality Optimization.

```
REM A BASIC program to find out variable combinations leading
REM to maximal Dsp and Fbr=5.03 simultaneously
REM This program is saved under the file name "optimum 1.bas"
CLS

REM Title and date in the optimization file
OPEN "optimum 1.dat" FOR OUTPUT AS #1
PRINT #1, "          #######     Optimization Results     ####### "
PRINT #1, "                      "; DATE$, TIME$
PRINT #1, "                      "
PRINT #1, "                      "
PRINT #1, "                      "
PRINT #1, "No.", "X1", "X2", "X3", "X4", "Dsp", "Fbr"

LET Dsp0 = 5
LET Fbr0 = 5.03

REM start scanning
FOR X1 = -1 TO 1.05 STEP .05
  FOR X2 = -1 TO 1.05 STEP .05
    FOR X3 = -1 TO 1.05 STEP .05
X4 = (10.05 - Fbr0 + 1.57*X1 + .75*X3 + .56*X1*X2 + 1.75*X1*X3
    + 1.21*X2*X3)/4.08
          IF X4 < -1 OR X4 > 1 THEN 2000
Dsp = 3.01 + 1.42*X1 - 1.05*X3 + .31*X4 - .51*X1*X3 + .9*X1*X4
    + .13*X2*X3 - .3*X3*X4
      IF Dsp < Dsp0 THEN GOTO 2000
          i = i + 1
          LET Dsp0 = Dsp
          PRINT #1, i; ":",
          PRINT #1, USING "##.###"; X1;
          PRINT #1, ",";
          PRINT #1, USING"##.###"; X2;
          PRINT #1, ",";
          PRINT #1, USING"##.###"; X3;
          PRINT #1, ",";
          PRINT #1, USING"##.###"; X4;
          PRINT #1, ",";
          PRINT #1, USING"##.###"; Dsp0;
          PRINT #1, USING"##.###"; Fbr0
2000 NEXT X3
  NEXT X2
NEXT X1

CLOSE #1
BEEP: BEEP
END
```

TABLE 3.31. Optimization Result of the Program in Table 3.30.

| | | ####### Optimization Results ####### | | | | |
| | | 05-02-1996 14:19:39 | | | | |
No.	$X1$	$X2$	$X3$	$X4$	D_{sp}	F_{br}
1:	0.150,	0.150,	-1.000,	0.999,	5.074	5.030
2:	0.200,	0.150,	-1.000,	0.997,	5.214	5.030
3:	0.250,	0.150,	-1.000,	0.996,	5.355	5.030
4:	0.300,	0.150,	-1.000,	0.995,	5.495	5.030
5:	0.350,	0.150,	-1.000,	0.994,	5.635	5.030
6:	0.400,	0.150,	-1.000,	0.993,	5.775	5.030
7:	0.450,	0.150,	-1.000,	0.991,	5.915	5.030
8:	0.500,	0.150,	-1.000,	0.990,	6.055	5.030
9:	0.550,	0.150,	-1.000,	0.989,	6.195	5.030
10:	0.600,	0.100,	-1.000,	0.999,	6.353	5.030
11:	0.650,	0.100,	-1.000,	0.997,	6.493	5.030
12:	0.700,	0.100,	-1.000,	0.996,	6.633	5.030
13:	0.750,	0.100,	-1.000,	0.994,	6.772	5.030
14:	0.800,	0.100,	-1.000,	0.993,	6.911	5.030
15:	0.850,	0.100,	-1.000,	0.991,	7.050	5.030
16:	0.900,	0.050,	-1.000,	0.998,	7.208	5.030
17:	0.950,	0.050,	-1.000,	0.996,	7.347	5.030
18:	1.000,	0.050,	-1.000,	0.994,	7.485	5.030

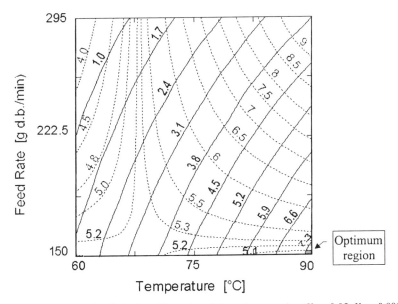

Figure 3.40 Graphical optimization: illustration of the optimum region ($X_2 = 0.05$, $X_4 = 0.99$).

CHAPTER 4

Food Recipe Modeling and Optimization

4.1 INTRODUCTION

4.1.1 GENERAL REMARKS ON RECIPE VARIABLES

A s discussed in Chapter 2, two types of questions exist in the food product design, namely, those raised in process modeling and those asked in recipe modeling. Their essential differences were already explained in Section 2.5. The handling of a process problem, that is, the modeling of the food process system and application of the model built for the purpose of product development was introduced in the last chapter. This chapter aims to set forth the techniques of how to solve a recipe or mixture problem in food product design.

A mixture problem is mainly concerned with recipe development research, such as formulation modeling and optimization. The food raw materials (mixture components or variables) in a mixture system are dependent on each other. This leads to the fundamental difference of the statistical experimental methodology applied in a food mixture system from that of a food process system. In a mixture system a specific relationship exists between all n ingredients [Equations (4.1) and (4.2)]. This means that no component variable can be changed without changing simultaneously any of the other component variables.

$$\sum_{j=1}^{n} x_j = 1 = 100\%$$

(4.1)

or in another form

$$X_j = 1.0 - \sum_{i=1}^{j-1} x_i - \sum_{i=j+1}^{n} x_i$$

(4.2)

125

4.1.2 MODEL FORMS

The problem of recipe development actually focuses on the modeling of a mixture system based on some limited experiments. Let us assume that there is a recipe in which a functional relationship exists between the n component X_1, X_2, X_3, ..., X_n, and a food quality index Z (response) [Equation (4.3)]. The n components will build a region called a simplex region, in which the food quality index Z would be related in a continuous fashion to mixtures comprised of X_1, X_2, X_3, ..., X_n and considered to be a response surface. Equation (4.3) exactly describes the response surface:

$$Z = f(X_1, X_2, X_3, \ldots, X_n) \tag{4.3}$$

Next, it can be thought that the fundamental consideration of recipe modeling is to determine the most suitable mathematical equation that can describe the surface in the simplex region with highest accuracy.

Similarly, as described in Section 3.3, a Taylor expansion equation is simple and generally suitable to represent the response surface $f(X_1, X_2, X_3, \ldots, X_n)$ approximately. Normally, a Taylor expansion equation of the first degree [Equation (4.4)] or of the second degree [Equation (4.5)] are the kind of models believed to well represent the surface.

$$Z = \beta_0 + \sum_{j=1}^{n} \beta_j \cdot X_j \tag{4.4}$$

$$Z = \beta_0 + \sum_{j=1}^{n} \beta_j \cdot X_j + \sum_{j<k=2}^{n} \beta_{jk} \cdot X_j \cdot X_k + \sum_{j=1}^{n} \beta_{jj} \cdot X_j^2 \tag{4.5}$$

The polynomial models used for recipe study in food product design appear to be the same as Equations (3.5) and (3.7). However, they are in fact different and probably even unfamiliar. They can, however, be transformed into the so-called Scheffé polynomial because of the specific relationship between all the n ingredients in a mixture system described by Equations (4.1) and (4.2). To give a simple example, a mixture with only two components X_1 and X_2 will be considered. Here the Taylor expansion equations of the first or second degree for these two variables are as given by Equation (4.6) and Equation (4.7), respectively:

$$Z = \beta_1 \cdot X_1 + \beta_2 \cdot X_2 \tag{4.6}$$

$$Z = \beta_1 \cdot X_1 + \beta_2 \cdot X_2 + \beta_{12} \cdot X_1 \cdot X_2 + \beta_{11} \cdot X_1^2 + \beta_{22} \cdot X_2^2 \tag{4.7}$$

For the two-component mixture system the following conversions can be made:

$$X_1 + X_2 = 1$$
$$X_1 = 1 - X_2$$
$$X_1^2 = X_1 \cdot (1 - X_2) = X_1 - X_1 \cdot X_2$$
$$X_2^2 = X_2 \cdot (1 - X_1) = X_2 - X_1 \cdot X_2$$

Thus the Taylor polynomial of the first degree [Equation (4.6)] can be transformed to:

$$
\begin{aligned}
Z &= \beta_0 + \beta_1 \cdot X_1 + \beta_2 \cdot X_2 \\
&= \beta_0 \cdot (X_1 + X_2) + \beta_1 \cdot X_1 + \beta_2 \cdot X_2 \\
&= (\beta_0 + \beta_1) \cdot X_1 + (\beta_0 + \beta_2) \cdot X_2
\end{aligned}
\tag{4.8}
$$

If we define:

$$(\beta_0 + \beta_1) = B_1$$
$$(\beta_0 + \beta_2) = B_2$$

Equation (4.6) will be transformed into a Scheffé polynomial of the first degree [Equation (4.9)].

$$Z = B_1 \cdot X_1 + B_2 \cdot X_2 \tag{4.9}$$

Reasoning in a similar way, a general model form for a mixture system with n components can be seen as:

$$Z = \sum_{j=1}^{n} B_j \cdot X_j \tag{4.10}$$

For a Taylor polynomial of the second degree in a mixture system with two components X_1 and X_2, the transformation is as given below:

$$
\begin{aligned}
Z &= \beta_0 + \beta_1 \cdot X_1 + \beta_2 \cdot X_2 + \beta_{12} \cdot X_1 \cdot X_2 + \beta_{11} \cdot X_1^2 + \beta_{22} \cdot X_2^2 \\
&= \beta_0 \cdot (X_1 + X_2) + \beta_1 \cdot X_1 + \beta_2 \cdot X_2 + \beta_{12} \cdot X_1 \cdot X_2 \\
&\quad + \beta_{11} \cdot (X_1 - X_1 \cdot X_2) + \beta_{22} \cdot (X_2 - X_1 \cdot X_2) \\
&= (\beta_0 + \beta_1 + \beta_{11}) \cdot X_1 + (\beta_0 + \beta_2 + \beta_{22}) \cdot X_2 \\
&\quad + (\beta_{12} - \beta_{11} - \beta_{22}) \cdot X_1 \cdot X_2
\end{aligned}
\tag{4.11}
$$

If we then define:

$$\beta_0 + \beta_1 + \beta_{11} = B_1$$
$$\beta_0 + \beta_2 + \beta_{22} = B_2$$
$$\beta_{12} - \beta_{11} - \beta_{22} = B_{12}$$

Equation (4.11) can be rewritten as Equation (4.12). Reasoning in a similar way, the general form of a second-degree mixture model for n variables can be obtained [Equation (4.13)]:

$$Z = B_1 \cdot X_1 + B_2 \cdot X_2 + B_{12} \cdot X_1 \cdot X_2 \tag{4.12}$$

$$Z = \sum_{j=1}^{n} B_j \cdot X_j + \sum_{j<k=2}^{n} B_{jk} \cdot X_j \cdot X_k \tag{4.13}$$

A model in the form of Equation (4.9) or Equation (4.12) is truly identical with Equation (4.6) or Equation (4.7) for a mixture system. The former two equations are normally used in mixture modeling and known as the Scheffé polynomial. The obvious difference between an original Taylor expansion equation and a Scheffé polynomial is that there are no constants and pure quadratic terms in the latter equation. However, the Scheffé polynomial model is derived from the Taylor expansion equation and is equivalent to it, having the same degree as the polynomial and the same number of components. The number of coefficients in a mixture model is smaller than in an ordinary polynomial. In addition, it must be noted that terms such as "$X_1 \cdot X_2$" in the quadratic Scheffé model cannot be thought of as the interaction effect between the variables X_1 and X_2. This is because the ingredient variables X_1 and X_2 are not independent of each other, that is, one variable cannot be changed without influencing the other. In fact, they describe only the nonlinear mixture effect of X_1 and X_2.

For an extremely complex mixture system, some cubic equation models can be used. They are usually used in the form of a fully cubic model [Equation (4.14)] or a specific cubic model [Equation (4.15)]. The general form of a specific cubic model for a mixture system with n components is given in Equation (4.16).

$$Z = B_1 \cdot X_1 + B_2 \cdot X_2 + B_{12} \cdot X_1 \cdot X_2 + B_{13} \cdot X_1 \cdot X_3 + B_{23} \cdot X_2 \cdot X_3$$
$$+ B_{12} \cdot X_1 \cdot X_2 \cdot (X_1 - X_2) + B_{13} \cdot X_1 \cdot X_3 \cdot (X_1 - X_3)$$
$$+ B_{23} \cdot X_2 \cdot X_3 \cdot (X_2 - X_3) + B_{123} \cdot X_1 \cdot X_2 \cdot X_3 \tag{4.14}$$

$$Z = B_1 \cdot X_1 + B_2 \cdot X_2 + B_{12} \cdot X_1 \cdot X_2 + B_{13} \cdot X_1 \cdot X_3$$
$$+ B_{23} \cdot X_2 \cdot X_3 + B_{123} \cdot X_1 \cdot X_2 \cdot X_3 \tag{4.15}$$

$$Z = \sum_{j=1}^{n} B_j \cdot X_j + \sum_{j<k=2}^{n} B_{jk} \cdot X_j \cdot X_k + \sum_{\substack{j<k=2 \\ i>2}}^{n} B_{jk} \cdot X_j \cdot X_k \cdot X_i \tag{4.16}$$

It is more convenient to establish a low-degree polynomial equation than a higher degree one, for it contains fewer terms and therefore requires fewer trials (observations) to permit the estimation of the coefficients B in the equation. As explained in Chapter 1, the selection of the equation depends mainly on the complexity of the studied mixture system. For a complicated system, a polynomial of the second or even third degree is required to describe the response surface. In general, the relationship between the food quality and the recipe components in a wide simplex region can only be modeled with a polynomial of higher degrees. Appropriate transformations of the ingredient variables can sometimes be considered, if the data fit is not adequate. Fortunately, a second-degree Scheffé polynomial will be adequate in most cases of food recipe development.

4.1.3 STEPS OF RECIPE DESIGN

Similar to studying a process problem, the general experimentation principle of response surface methodology (RSM) is also suitable for recipe or formulation study. By way of modeling, the food properties can be predicted in various formulations, and the blending properties of the components as well as the effects of the separate components on the food quality can be established. The five main steps and considerations related to the exploration of the response surface over a simplex region for recipe design are given below. They will be explained in detail in the following part of this chapter.

(1) Step I: Ingredient screening: Identify and select the important and meaningful recipe components (mixture variables) that have significant effects on the food product qualities (responses); define and set the simplex region of interest for study.
(2) Step II: Mixture experimental design: Design with care, a suitable mixture experimental design for collecting observations corresponding to the form of the mathematical model to be fitted. This is an essential step for the formulation design, because the properties of the polynomials used to estimate the food quality depend to a large extent on the pattern of the design.
(3) Step III: Model building: Build the model, which approximately describes the relationship between the components (variables) and the food quality indices (responses) over the simplex region.

(4) Step IV: Model test: Test the suitability of the model for representing the response surface in the simplex region.
(5) Step V: Model application: Analyze the effects of the components both separately and jointly on the food quality (responses); perform optimization and prediction based on the significant model.

4.2 TARGET DEFINITION AND FOOD INGREDIENT SCREENING

4.2.1 RESPONSE DEFINITION

As the first step, the objective food quality characteristics of the targeted food product must be identified and selected. The general information about the identification and selection of food quality indices supplied in Section 3.2.1 can also be applied to a mixture system. In general, those quality indices that are specific and important for the food product to be developed must be selected as responses. Furthermore, they must be determined and presented numerically to make modeling possible.

4.2.2 INGREDIENT SCREENING

As important component variables, the raw materials in the recipe that significantly affect the desired quality of the food products must be identified. During this step, cost and availability of the raw materials must be considered. Also, attempts should be made to minimize the number of raw materials or components to be studied to simplify the subsequent research. If there are too many ingredients in the recipe, some of them, which are believed to be relatively unimportant, should be held constant during the study. Normally, the study of a mixture with fewer than five ingredients or components is reasonable with respect to the extent of trials. Next, the variation range of the chosen ingredients must be defined, in preparation for the mixture experimental design.

4.3 DESIGN OF MIXTURE EXPERIMENTAL PLAN

To approximate the functional relationship as a Scheffé polynomial, such as Equations (4.10) and (4.13), or to any other form of a model or equation, some preselected experimental runs are performed at various predetermined combinations of the proportions of the q components. Such a set of combinations (or recipes) of the proportions is referred to as the mixture experimental design and should cover the range of interest to the food product developer. Once the n observations are collected, the parameters in the model can be estimated by the method of least squares.

The mixture experimental designs are unique because they consider those constraints that make the usual computer programs unsuitable. They are typically performed whenever the response is a function concerning the proportions of the variables and not that of their total quantities. For this reason, mixture experiments can be used in a wide range of application in food product development.

4.3.1 SELECTION OF EXPERIMENTAL RANGE

As mentioned above, a mixture experimental design including less than five ingredients proved to be reasonable in practice. After the selection of ingredients the varying ranges of interest for every ingredient must be determined. Some knowledge about the product and the ingredients used is usually needed so that reasonable ranges of the ingredients can be determined. In fact, if there is truly an optimal formulation, then there must be ingredient levels that are higher and lower than the optimal value. It is at this planning stage that some of the artistry of the process comes into play. In principle, the range of an ingredient consists of two extreme values (levels), and the ranges of all ingredients make up the studied simplex region. According to the extreme values of each ingredient to be studied, a mixture design could be set up. For the building of the design some basic principles should be considered:

- The selected experiments should distribute evenly and regularly over the simplex region to get well-balanced information from the supporting points.
- The model to be built based on the supporting points should be able to describe the real observations (response) as precisely as possible.
- Enough trials should be used to ensure detection of any lack of fit of the model.
- It should be possible to estimate the error variance during realization of the experiments.

In an experimental design the range intended by the food formulation developers should be defined with respect to the proportions used for each of the components. If the interest is targeted on all the values of X_i ranging from 0 to 1.0, the design may cover the entire simplex factor room; otherwise, the design might cover only a subportion or a smaller subspace within the simplex. This latter situation comes up in practice when additional constraints in the form of upper and/or lower bounds are placed on the component proportions or when the interest is only in a group of mixtures that are located in some small regions inside the simplex. The pattern of the experimental design decides to a large extent the properties of the polynomials used to estimate the response function.

4.3.2 SELECTION OF SUITABLE MODELS

When selecting a mixture experimental design, consideration must be given to the type of model to ensure a proper fit, because the model form and properties depend largely on the design matrix and the experimental design. To fit a polynomial of the first or second degree, different mixture experimental designs should be made and adopted. As discussed above, the selection of the equation form depends mainly on the complexity of the mixture system and the extent of the simplex region to be studied. The linear mixture model [Equation (4.11)] is the simplest form of recipe modeling. In this case it is assumed that the effects of the mixture components (variables) on the food product quality (response) are additive. However, in most practical cases the quality indices of the food from some formulations are better than that of the added effect of all ingredients separately or singly in the recipe. This means that the response can be better described with a model of second [Equation (4.13)] or even of the third degree [Equation (4.16)] than with a linear model. Generally, the quadratic model [Equation (4.13)] proves to be suitable and is thus used in most cases.

4.3.3 TYPICAL MIXTURE EXPERIMENTAL DESIGNS

Similar to the various types of experimental designs for process modeling, there are several kinds of typical mixture experimental designs supporting the building of models of the first or second degree. According to the basic principles of statistical experimental methodology, some typical mixture experimental designs, which support the building of such models, are discussed in the following sections.

4.3.3.1 Simplex Lattice Design

Simplex lattice designs were introduced by Scheffé in the 1960s in developing the mixture experimental methods. These designs have been credited by many statisticians to be the foundation of the theory of experimental designs for mixtures. In this method the experimental combinations to be tested distribute evenly and regularly in the simplex region, that is, in the experimental space.

To enable a polynomial equation to represent the response surface over the entire simplex region, a natural choice for a mixture design should be one whose points are spread uniformly over the whole simplex factor space. An orderly arrangement consisting of an evenly spaced distribution of points on a simplex is known as a lattice. A specific lattice may have a special correspondence to a specific polynomial equation. In principle, to support the building of a polynomial model of degree m in q components over the simplex, the lattice, usually referred to as a $\{q, m\}$ simplex lattice, consists of points whose coordinates are defined by the following combinations of the component proportions. The proportions of each component take the $m + 1$ equally spaced values from 0 to 1 [Equation (4.16)]:

$$X_i = 0, \quad 1/m, \quad 2/m, \dots, 1 \tag{4.17}$$

The $\{q, m\}$ consists of all possible combinations or mixtures of the components where the proportions for each component are used. The number of design points in the $\{q, m\}$ simplex lattice is

$$\binom{q + m - 1}{m} = \frac{(q + m - 1)!}{m!(q - 1)!} \tag{4.18}$$

where $m!$ is the m factorial and can be calculated with Equation (4.19):

$$m! = m \cdot (m - 1) \cdot (m - 2) \cdot \dots \cdot 2 \cdot 1 \tag{4.19}$$

4.3.3.1.1 Necessary Number of Trials for Different Models

As discussed above, the building of models of different degrees requires different numbers of experiments in the mixture design. Table 4.1 shows the minimal numbers of necessary trials for building mathematical models of different degrees—linear, quadratic, or cubic—in terms of a simplex lattice mixture design. In general, a simplex lattice design for building a quadratic model can also be used to fit a linear model, thus getting an even more precise fit. Hence, it is suggested that a simplex lattice design should be used, regardless of whether a linear or a quadratic model is to be fitted. Of course, if the experiments are expensive in terms of raw materials or the time consumed, one can consider making a design supporting the building of a linear model. However, such cases are rare in the food industry.

As shown in Table 4.1, fewer supporting points are needed to build models of lower degrees, meaning fewer trials and lower cost. An obvious disadvantage of using a simplex lattice design to study a mixture system is that the number of experiments will rise rapidly if more than five components are to be studied and a cubic model—full or specific—is to be built. Fortunately, in food mixture

TABLE 4.1. Necessary Number of Trials for Building of Different Models.

m	Linear Model	Quadratic Model	Specific Cubic Model	Full Cubic Model
2	2	3	—	—
3	3	6	7	10
4	4	10	14	20
5	5	15	25	35
6	6	21	41	56

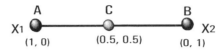

Figure 4.1 Geometrical description of the simplex lattice design in a mixture system with two components X_1 and X_2.

systems most of the response surfaces over the simplex region can be described precisely enough with a quadratic model. Some concrete examples of simplex lattice designs for systems with two to four mixture components and the building of a corresponding quadratic and cubic model are shown in the following sections.

4.3.3.1.2 Mixture with Two Components

For a binary blend with two components, X_1 and X_2, the supporting points to be tested should be located at the mixtures A, B, and C whose coordinates would be $(1, 0)$, $(1/2, 1/2)$, and $(0, 1)$. Geometrically, these points are illustrated in Figure 4.1. This design will support the building of a quadratic as well as a linear model.

4.3.3.1.3 Mixture with Three Components

This section shows what a simplex lattice design would be for a mixture system with three components X_1, X_2, and X_3. As shown in Table 4.1, to build a mathematical model of the second or third degree requires a different number of supporting points. More trials must be performed for building a cubic model than that for the linear or quadratic models. The supporting points in the simplex lattice design for building a quadratic and cubic model are as follows:

(1) Building of a linear or a quadratic model: These supporting points— or the whole simplex lattice design—are (is) geometrically illustrated in Figure 4.2. It is obvious that the experimental points consist of all three

TABLE 4.2. A {3, 2} Simplex Lattice Design.

Number of Trials	Recipes			Coordinates	Points in Figure 4.2
	X_1	X_2	X_3		
1	1	0	0	(1, 0, 0)	A
2	0	1	0	(0, 1, 0)	B
3	0	0	1	(0, 0, 1)	C
4	1/2	1/2	0	(1/2, 1/2, 0)	D
5	1/2	0	1/2	(1/2, 0, 1/2)	E
6	0	1/2	1/2	(0, 1/2, 1/2)	F

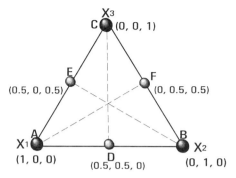

Figure 4.2 Geometrical description of the simplex lattice design for building of a quadratic model with three components X_1, X_2, and X_3.

vertices (A, B, and C) and the midpoint of the three sides of the triangle (D, E, and F).

(2) Building of a cubic model: Figure 4.3 shows geometrically these support-ing points in Table 4.3. The experimental points consist of all three vertices (A, B, and C) and the one-third points on the three sides of the triangle (D, E, F, G, H, and I).

4.3.3.1.4 *Mixture with Four Components*

Similar to mixture systems with three components, the simplex lattice design for quadratic and cubic model building in a system with four components X_1, X_2, X_3, and X_4 is as follows:

(1) Building of a linear or a quadratic model: The supporting points are illus-trated in Figure 4.4. The experimental points consist of all three vertices

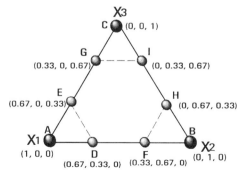

Figure 4.3 Geometrical description of the simplex lattice design for building a cubic model with three components X_1, X_2, and X_3.

TABLE 4.3. A {3, 3} Simplex-Lattice Design.

Number of Trials	Recipes			Coordinates	Points in Figure 4.3
	X_1	X_2	X_3		
1	1	0	0	(1, 0, 0)	A
2	0	1	0	(0, 1, 0)	B
3	0	0	1	(0, 0, 1)	C
4	2/3	1/3	0	(2/3, 1/3, 0)	D
5	2/3	0	1/3	(2/3, 0, 1/3)	E
6	1/3	2/3	0	(1/3, 2/3, 0)	F
7	1/3	0	2/3	(1/3, 0, 2/3)	G
8	0	2/3	1/3	(0, 2/3, 1/3)	H
9	0	1/3	2/3	(0, 1/3, 2/3)	I

(A, B, C, and D) and the midpoint of the six sides of the triangle (E, F, G, H, I, and J). The trials are shown in Table 4.4.

(2) Building of a cubic model: These supporting points are illustrated in Figure 4.5. Obviously, the experimental points consist of all four vertex points (A, B, C, and D) and 12 one-third points of all six edges of the tetrahedron (E, F, G, H, I, J, K, L, M, N, O, and P).

In addition to the typical simplex lattice mixture design, the central point of the studied region can be included in the experimental design to achieve a more

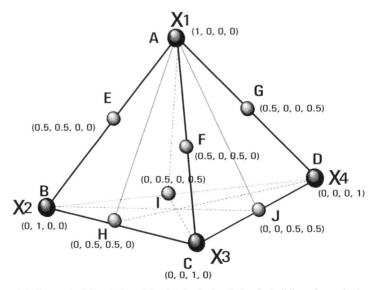

Figure 4.4 Geometrical description of the simplex lattice design for building of a quadratic model with four components X_1, X_2, X_3, and X_4.

TABLE 4.4. A {4, 2} Simplex-Lattice Design.

Number of Trials	Recipes				Coordinates	Points in Figure 4.4
	X_1	X_2	X_3	X_4		
1	1	0	0	0	(1, 0, 0, 0)	A
2	0	1	0	0	(0, 1, 0, 0)	B
3	0	0	1	0	(0, 0, 1, 0)	C
4	0	0	0	1	(0, 0, 0, 1)	D
5	1/2	1/2	0	0	(1/2, 1/2, 0, 0)	E
6	1/2	0	1/2	0	(1/2, 0, 1/2, 0)	F
7	1/2	0	0	1/2	(1/2, 0, 0, 1/2)	G
8	0	1/2	1/2	0	(0, 1/2, 1/2, 0)	H
9	0	1/2	0	1/2	(0, 1/2, 0, 1/2)	I
10	0	0	1/2	1/2	(0, 0, 1/2, 1/2)	J

uniform distribution of the experimental points (supporting points) and get one more degree of freedom for the observations. This forms the so-called simplex centroid design that will be discussed below.

4.3.3.2 Simplex Centroid Design

As stated above, the simplex centroid design is quite similar to the simplex lattice design, except for additional central points of the simplex region to be explored. This design is suitable for building quadratic and cubic models with

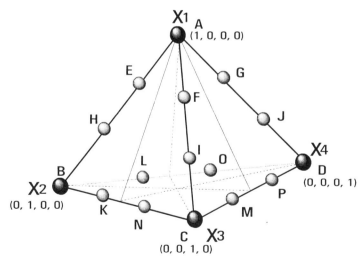

Figure 4.5 Geometrical description of the simplex lattice design for building of a cubic model with four components X_1, X_2, X_3, and X_4.

TABLE 4.5. A {4, 3} Simplex-Lattice Design.

Number of Trials	Recipes				Coordinates	Points in Figure 4.5
	X_1	X_2	X_3	X_4		
1	1	0	0	0	(1, 0, 0, 0)	A
2	0	1	0	0	(0, 1, 0, 0)	B
3	0	0	1	0	(0, 0, 1, 0)	C
4	0	0	0	1	(0, 0, 0, 1)	D
5	2/3	1/3	0	0	(1/2, 1/2, 0, 0)	E
6	2/3	0	1/3	0	(1/2, 0, 1/2, 0)	F
7	2/3	0	0	1/3	(1/2, 0, 0, 1/2)	G
8	1/3	2/3	0	0	(1/2, 1/2, 0, 0)	H
9	1/3	0	2/3	0	(1/2, 0, 1/2, 0)	I
10	1/3	0	0	2/3	(1/2, 0, 0, 1/2)	J
11	0	2/3	1/3	0	(0, 1/2, 1/2, 0)	K
12	0	2/3	0	1/3	(0, 1/2, 0, 1/2)	L
13	0	0	2/3	1/3	(0, 0, 1/2, 1/2)	M
14	0	1/3	2/3	0	(0, 1/2, 1/2, 0)	N
15	0	1/3	0	2/3	(0, 1/2, 0, 1/2)	O
16	0	0	1/3	2/3	(0, 0, 1/2, 1/2)	P

more than four mixture components. In a mixture system with q components, the number of distinct points in a simplex centroid design is $2^q - 1$. These points contain every (nonempty) combination of the q components, but only with mixtures in which the components are presented in equal proportions. Such mixtures are located at the centroid of the $(q - 1)$ dimensional simplex region and at the centroids of all the lower dimensional simplexes contained within the $(q - 1)$-dimensional simplex. More exactly, these $2^q - 1$ trials can be divided into several types of subgroups:

- $\binom{q}{1}$ permutations $(1, 0, 0, \ldots, 0)$, q pure components
- $\binom{q}{2}$ permutations $(0.5, 0.5, 0, \ldots, 0)$, binary mixtures
- $\binom{q}{3}$ permutations $(0.333, 0.333, 0.333, 0, \ldots, 0)$, triple mixtures
- ....
- ....
- $\binom{q}{q-1}$ permutations $[1/(q - 1), 1/(q - 1), 1/(q - 1), \ldots, 1/(q - 1), 0]$, one mixture with $q - 1$ components
- one central point $(1/q), 1/q, 1/q, \ldots, 1/q, \ldots, 1/q)$

For a better understanding of the structure of simplex centroid designs, two mixture systems with three and four components are presented below as examples. In Figure 4.6 the design is illustrated for a mixture system with three components X_1, X_2, and X_3, and the trials are shown in Table 4.6.

Figure 4.7 shows the distribution of the observations for a mixture system with four components X_1, X_2, X_3, and X_4. The corresponding mixtures (recipes) to be tested are listed in Table 4.7.

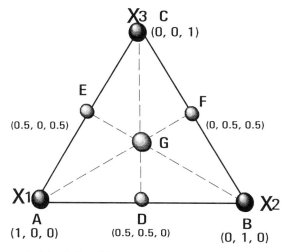

Figure 4.6 Geometrical description of the simplex centroid design for a mixture system with three components X_1, X_2, and X_3.

4.3.3.3 Simplex Axial Design

The simplex lattice and the simplex centroid designs are boundary designs in which the points are positioned on the boundaries, such as vertices, edges, and faces of the simplex factor space, with the exception of the overall centroid. Axial designs, on the other hand, are designs consisting mainly of complete mixtures or blends of q components where most of the points are positioned inside the simplex. Figure 4.8 shows the observation points for a simplex axial mixture system with three components X_1, X_2, and X_3. The corresponding mixtures (recipes) to be tested are listed in Table 4.8.

In general, axial designs have been recommended for use when component effects are to be measured in screening experiments, particularly when models of the first degree are to be fitted.

TABLE 4.6. **A Simplex Centroid Design with Three Components.**

Number of Trials	Recipes X_1	X_2	X_3	Coordinates	Points in Figure 4.6
1	1	0	0	(1, 0, 0)	A
2	0	1	0	(0, 1, 0)	B
3	0	0	1	(0, 0, 1)	C
4	1/2	1/2	0	(1/2, 1/2, 0)	D
5	1/2	0	1/2	(1/2, 0, 1/2)	E
6	0	1/2	1/2	(0, 1/2, 1/2)	F
7	1/3	1/3	1/3	(1/3, 1/3, 1/3)	G

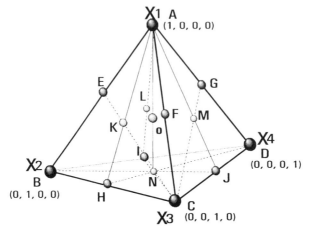

Figure 4.7 Geometrical description of the simplex centroid design for a mixture system with four components X_1, X_2, X_3, and X_4.

4.3.3.4 Constrained Mixture Design

In most cases of food product design it is impossible or not necessary to model and to explore the entire simplex region. Often, there are certain additional restrictions placed on the component proportions. For example, the baking powder may not be varied from 0% to 100%, but reasonably from 0.5% to 1.5%, which limits the desired mixtures to a subregion of the simplex. Furthermore,

TABLE 4.7. A Simplex Centroid Design with Four Components.

Number of Trials	Recipes				Coordinates	Points in Figure 4.7
	X_1	X_2	X_3	X_4		
1	1	0	0	0	(1, 0, 0, 0)	A
2	0	1	0	0	(0, 1, 0, 0)	B
3	0	0	1	0	(0, 0, 1, 0)	C
4	0	0	0	1	(0, 0, 0, 1)	D
5	1/2	1/2	0	0	(1/2, 1/2, 0, 0)	E
6	1/2	0	1/2	0	(1/2, 0, 1/2, 0)	F
7	1/2	0	0	1/2	(1/2, 0, 0, 1/2)	G
8	0	1/2	1/2	0	(0, 1/2, 1/2, 0)	H
9	0	1/2	0	1/2	(0, 1/2, 0, 1/2)	I
10	0	0	1/2	1/2	(0, 0, 1/2, 1/2)	J
11	1/3	1/3	1/3	0	(1/3, 1/3, 1/3, 0)	K
12	1/3	1/3	0	1/3	(1/3, 1/3, 0, 1/3)	L
13	1/3	0	1/3	1/3	(1/3, 0, 1/3, 1/3)	M
14	0	1/3	1/3	1/3	(0, 1/3, 1/3, 1/3)	N
15	1/4	1/4	1/4	1/4	(0, 1/4, 1/4, 0)	O

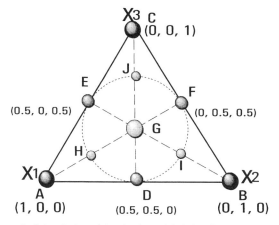

Figure 4.8 Geometrical description of the simplex axial design for a mixture system with three components X_1, X_2, and X_3.

if only a subregion of an entire simplex is to be modeled, the precision of the model can be expected to increase, whereas time-consuming and experimental cost is reduced.

In general, there are three kinds of restrictions in the mixture study, namely, the lower-bound restrictions, the upper-bound restrictions, and the restrictions with both lower and upper bounds. Constructing a suitable mixture experimental design for a constrained subsimplex region is important for effective and successful response modeling and optimization.

On the basis of those common mixture designs presented above, it is easy to design the experiments for subsimplex problems with only lower- or upper-bound restrictions. Some rules can help the designer to select suitable blends for a reasonable mixture design. Usually, the mixtures to be tested should include

TABLE 4.8. A Simplex Axial Design for Three Components.

Number of Trials	Recipes			Coordinates	Points in Figure 4.8
	X_1	X_2	X_3		
1	1	0	0	(1, 0, 0)	A
2	0	1	0	(0, 1, 0)	B
3	0	0	1	(0, 0, 1)	C
4	1/2	1/2	0	(1/2, 1/2, 0)	D
5	1/2	0	1/2	(1/2, 0, 1/2)	E
6	0	1/2	1/2	(0, 1/2, 1/2)	F
7	1/3	1/3	1/3	(1/3, 1/3, 1/3)	G
8	2/3	1/6	1/6	(2/3, 1/6, 1/6)	H
9	1/6	2/3	1/6	(1/6, 2/3, 1/6)	I
10	1/6	1/6	2/3	(1/6, 1/6, 2/3)	J

the subsets of the extreme vertices, the edge centroids, the constraint designes centroids, and the overall centroid.

There is a practical way for easy designing of a mixture experimental design with restrictions. First, the upper and lower levels of all components but one are combined with each other. Thus, there would be $q \cdot 2^{(q-1)}$ possible combinations of components created with one blank level. Then all these combinations are completed by replacing the blank levels of one component with its restricted levels, and the sum of the components is checked in each combination. If it is 100% or 1.0%, then this mixture represents a vertex and should be selected in the mixture experimental design. The centroid of each two-dimensional (2D) surface and of any three and more dimensional arrangement, including the overall centroid, can be calculated from the mixture of the respective vertices.

4.4 REALIZATION OF MIXTURE EXPERIMENTAL DESIGNS

The trials in a mixture experimental design must be performed according to the similar instructions that have been discussed in Section 3.5. While collecting observations, an attempt should be made to collect replication observations at several points of the whole design. This will permit an estimate of the observation variance and may improve the accuracy of the estimated model.

4.5 MODEL BUILDING

The objective of this stage is to fit a proposed model that describes the food quality (response surface) continuously over the simplex region and to further determine the effects of the individual components on the food quality. As already discussed at the beginning of this chapter, the simpler the fitted model form, the easier it is to understand the blending properties of components and to use the model for predictive purposes, particularly for those blends that were not used for fitting the model. However, it is also usually true that the simpler the form of the model, the less precise it will be. The high precision of a complete model should not be sacrificed while simplifying the model.

4.5.1 SPECIFIC REGRESSION TECHNIQUES

In mixture research the initially proposed model to which the observations in each of the data sets will be fitted is usually a Scheffé model. However, most of the standard computer regression programs cannot provide the correct analysis information of a fitted Scheffé-type mixture model because:

- The fitted model by the program should contain a constant term, but a Scheffé model does not.

- If a no-constant-term option is available in the program and has been selected in the regression procedure, then the regression sum of squares in the analysis of the variance (ANOVA) table cannot be corrected for the overall mean.

As a result of the latter cause stated above, the F-test and R^2 values in a standard software output are incorrect, because in such programs the unadjusted (in reference to the mean) regression sum of squares and the total sum of squares are used for computation. In general, the values of F and R^2 values are inflated, giving the wrong impression that the fitted model is better than it really is. The simplest way that still allows using the computer as an effective tool in the regression is to calculate the F and R^2 values in a spreadsheet such as Ms-Excel$^{®}$ or Lotus 1-2-3$^{®}$ manually, which may be a little bit laborious.

The inclusion of the constant term B_0 is necessary for many least square regression programs for the regression and total sum of squares in the ANOVA table to be calculated correctly. There is an easier way to obtain the correct regression results using the standard statistical packages. To do this, one linear term $B_i \cdot X_i$ in the model is first deleted, and the modified model is fitted by using a normal standard program. The multiplicated terms $X_i \cdot X_j$ ($j = 1, 2, \ldots, q, j \neq i$) are still calculated as usual. Once the model is fitted, one can easily reexpress the fitted equation with a canonical polynomial model of the Scheffé type, just by rewriting the constant term B_0 as $B_0 \cdot (X_1 + X_2 + \ldots X_q)$ in the model.

As an example, using an SAS macro to estimate the coefficients in a mixture model, with and without constant, is listed in Table 4.9. The statement NOINT behind the model form instructs SAS that no constant term (intercept) should be included in the model. The three terms $X_1 \cdot X_2$, $X_1 \cdot X_3$, and $X_1 \cdot X_4$ have been found coincidentally insignificant and not included in the model. Part of the running results of this macro is printed out in Table 4.10. The regression shows that the F-value (276.24) and R^2 (0.99742) from the regression without a constant term are not identical with that of the correct regression (F-value: 60.48; $R^2 : 0.986408$). The final regression model of Scheffé type rewritten from the coefficients in Table 4.10 should be as follows:

$$Y = 1.94 \cdot X_1 + 2.1 \cdot X_2 + 0.04 \cdot X_3 + 0.29 \cdot X_4$$
$$+ 6.29 \cdot X_2 \cdot X_3 + 5.64 \cdot X_2 \cdot X_4 + 2.47 \cdot X_3 \cdot X_4$$

4.6 VALIDITY OF ESTABLISHED MODELS

After the model has been established, its significance must be tested to decide whether it is workable. The basic theory of a significance test has been introduced briefly in Section 3.6.4 and will not be repeated here. However, as stated above, although the F- and R^2 values for a mixture model cannot be

TABLE 4.9. A Macro of Estimating the Coefficients without or with the Constant Term.

	data;				
	input dpoint x1 x2 x3 X4 y;				
cards;					
1	.180	.820	.000	.000	2.078
2	.180	.000	.820	.000	.450
3	.180	.000	.000	.820	.588
4	.180	.273	.273	.273	2.176
5	.300	.700	.000	.000	2.079
6	.300	.000	.700	.000	.572
7	.300	.000	.000	.700	.821
8	.300	.233	.233	.233	2.066
9	.240	.380	.380	.000	2.066
10	.240	.380	.000	.380	2.066
11	.240	.000	.380	.380	.827
12	.240	.253	.253	.253	2.077
run;					
prog glm;					
	model y = x1 x2 x3 X4 x2*x3 x2*x4 x3*x4/ noint;				
prog glm;					
	model y = x2 x3 X4 x2*x3 x2*x4 x3*x4;				
run;					

correctly calculated by a usual computer program, they can be calculated with the help of a table calculation program such as Ms-Excel®, Lotus 1-2-3®, and so on. Although these values are calculated, it must be noted that the degrees of freedom of the regression and the error are $(q - 1)$ and $(n - q)$, respectively, where q is the number of the terms that are estimated in the fitted Scheffé model, and n is the total number of observations.

The F-test is based on the hypothesis that the food quality index does not depend on the proportion of the components. The value of the F-ratio is calculated according to Equation (3.17) and is compared with the critical tabled value of $F_{(q-1,n-q)}^{\alpha}$. The hypothesis is rejected at the α level of significance if the value of the F-ratio exceeds the critical tabled value. The dependence of the response on the mixture components can thus be confirmed. In other words, the magnitude of the response does vary with the various combinations of components.

The R^2 value is used to examine whether the model explains a sufficient amount of the variation in the real response values. The value of $1 - R^2$ is the percentage ratio of the error variance estimate obtained from the analysis of the regression model to the error variance estimate obtained by using the model (Section 3.6.4.2). So a relatively high R^2 value is a sign that the regression model can be used with confidence for the purpose of predicting response values.

The criteria used for checking the model in a process system are also applicable to a mixture system. If a first-degree model proves to be inadequate, then

TABLE 4.10. Regression Results of the Macro in Table 4.9.

(model without the constant term)

SAS   16:33 Sunday, June 2, 1996 11
General Linear Models Procedure
Number of observations in data set = 12

Dependent Variable: Y

Source	DF	Sum of Squares	Mean Square	F Value	Pr > F
Model	7	32.64328893	4.66332699	**276.24**	0.0001
Error	5	0.08440707	0.01688141		
Uncorrected Total	12	32.72769600			

R-Square	C.V.	Root MSE	Y Mean
0.997421	8.726866	0.129928	1.48883333

Dependent Variable: Y

Source	DF	Type I SS	Mean Square	F Value	Pr > F
X1	1	25.71162050	25.71162050	1523.07	0.0001
X2	1	4.45372023	4.45372023	263.82	0.0001
X3	1	0.00007457	0.00007457	0.00	0.9496
X4	1	0.23405251	0.23405251	13.86	0.0137
X2*X3	1	1.32698429	1.32698429	78.61	0.0003
X2*X4	1	0.78601049	0.78601049	46.56	0.0010
X3*X4	1	0.13082632	0.13082632	7.75	0.0387

Source	DF	Type III SS	Mean Square	F Value	Pr > F
X1	1	0.18637504	0.18637504	11.04	0.0209
X2	1	1.57700433	1.57700433	93.42	0.0002
X3	1	0.00075783	0.00075783	0.04	0.8406
X4	1	0.03071879	0.03071879	1.82	0.2352
X2*X3	1	0.84432005	0.84432005	50.01	0.0009
X2*X4	1	0.67891756	0.67891756	40.22	0.0014
X3*X4	1	0.13082632	0.13082632	7.75	0.0387

Parameter	Estimate	T for H0: Parameter = 0	Pr > |T|	Std Error of Estimate
X1	1.937262067	3.32	0.0209	0.58304075
X2	2.102050560	9.67	0.0002	0.21748591
X3	0.046080153	0.21	0.8406	0.21748591
X4	0.293378846	1.35	0.2352	0.21748591
X2*X3	6.300943773	7.07	0.0009	0.89095642
X2*X4	5.650157740	6.34	0.0014	0.89095642
X3*X4	2.480273767	2.78	0.0387	0.89095642

(model with the constant term but without X_1)

General Linear Models Procedure
Dependent Variable: Y

Source	DF	Sum of Squares	Mean Square	F Value	Pr > F
Model	6	6.04490693	1.00748449	**60.48**	0.0002
Error	5	0.08329273	0.01665855		
Corrected Total	11	6.12819967			

R-Square	C.V.	Root MSE	Y Mean
0.986408	8.669069	0.129068	1.48883333

(continued)

145

TABLE 4.10. (Continued)

Dependent Variable: Y					
Source	DF	Type I SS	Mean Square	F Value	Pr > F
X2	1	3.72915974	3.72915974	223.86	0.0001
X3	1	0.03970441	0.03970441	2.38	0.1833
X4	1	0.03818326	0.03818326	2.29	0.1905
X2*X3	1	1.32382829	1.32382829	79.47	0.0003
X2*X4	1	0.78393899	0.78393899	47.06	0.0010
X3*X4	1	0.13009224	0.13009224	7.81	0.0382
Source	DF	Type III SS	Mean Square	F Value	Pr > F
X2	1	0.00070564	0.00070564	0.04	0.8451
X3	1	0.10162986	0.10162986	6.10	0.0565
X4	1	0.07686955	0.07686955	4.61	0.0844
Dependent Variable: Y					
Source	DF	Type III SS	Mean Square	F Value	Pr > F
X2*X3	1	0.84245923	0.84245923	50.57	0.0009
X2*X4	1	0.67724817	0.67724817	40.65	0.0014
X3*X4	1	0.13009224	0.13009224	7.81	0.0382

	T for H0: Pr > |T| Std Error of			
Parameter	Estimate	Parameter = 0		Estimate
Intercept	1.942442309	3.35	0.0202	0.57899972
X2	0.158139063	0.21	0.8451	0.76836133
X3	−1.897831343	−2.47	0.0565	0.76836133
X4	−1.650532651	−2.15	0.0844	0.76836133
X2*X3	6.293961794	7.11	0.0009	0.88505083
X2*X4	5.643175762	6.38	0.0014	0.88505083
X3*X4	2.473291789	2.79	0.0382	0.88505083

either additional experiments must be performed that might improve the fit, or a quadratic model must be adopted.

4.7 ANALYSIS OF COMPONENTS EFFECT

In a similar way as described in Section 3.7, screening out the unimportant components from a group of components by measuring and detecting the effects of the most important components is a further task of the food recipe design. The motive for wanting to detect the important components is to reveal and control them in the mixture system to work out the most reasonable recipe. To analyze the separate and joint effects of mixture variables, graphical and numerical approaches are usually used. They are discussed in the following sections.

4.7.1 RESPONSE TRACE PLOT

To estimate the effect of component X_i on the response values, a trace plot proves to be a useful tool. It is generated by using the fitted model along the directions described as follows [Equations (4.20) and (4.21)]:

$$X_i = S_i + \Delta_i \tag{4.20}$$

$$X_j = S_j - \frac{\Delta_i \cdot S_j}{1 - S_i} \tag{4.21}$$

where $j = 1, 2, \ldots, q$, $j \neq i$, $1 - S_i \geq 0$, and Δ_i is the increment that could be a positive or negative value. In general, the response trace plot is generated in four basic steps:

(1) Step I: Select a reference mixture S, which generally will be the centroid of the experimental region. [If the centroid point is selected as the start point, then for the case of no restrictions of the other $n - 1$ mixture components, the calculation of the coordinate values of the other $n - 1$ mixture variables can be simply obtained through $(1 - X_i)/(n - 1)$.]
(2) Step II: On the X_i-ray, the increment of component i increases by an amount Δ_i [Equation (4.20)] by moving away either from the reference mixture toward the vertex $X_i = 1$ ($\Delta_i \geq 0$) or from the vertex to $X_i = 0$ ($\Delta_i \neq 0$). Keep the other component proportions in the same ratio as that at the reference mixture S [Equation (4.21)]. Each Δ_i value corresponds to a blend (with a set of coordinates) on the X_i-ray. Choose some blends to make predictions of the response (only within the experimental region!).
(3) Step III: Put the coordinate values of each blend selected in step II into the regression model to obtain the predicted response values along the X_i-ray. Plot the predicted response values against X_i.
(4) Step IV: Repeat steps II and III on the other X_j-rays ($j = 1, 2, \ldots, q$, $j \neq i$) to plot the response trace for the other components. There will be a total of q plots.

This procedure of generating response trace plots is quite troublesome if it is performed manually. However, it can be done with great ease with the aid of a table calculation software. The following example shows this technique.

In a food extrusion study, the relationship between the in vitro protein efficiency ratio (DC-PER) of the extrudates and the four raw materials, namely, water (M), soybean flour (S), rice flour (R), and wheat flour (W) were researched and modeled. The raw materials were mixed prior to the extrusion according to an experimental design based on the simplex centroid mixture design. The moisture content was limited within a reasonable range of 0.18–0.30; namely, 18–30%, and soybean, rice, and wheat can thus only vary from 0 to 0.82. The model obtained is as follows [Equation (4.22)]:

$$\text{DC-PER} = 1.478 \cdot M + 2.079 \cdot S + 0.585 \cdot R + 0.761 \cdot W$$

$$+ 4.022 \cdot S \cdot R + 3.635 \cdot S \cdot W + 0.845 \cdot R \cdot W \tag{4.22}$$

To plot the effects of water, soybean, rice, and wheat on the protein efficiency ratio in a response trace plot, the steps listed above would be used. The following steps show concretely how to construct the response trace plots for water.

(1) As described in step I, select the central point at the surface of lower bound of water in the researched region (0.18, 0.273, 0.273, 0.273) as the reference mixture.
(2) Set the increment Δ_i of M at 0.02. Calculate the first simplex point in moving away from the reference mixture as follows:

$$M_1 = 0.18 + 0.02 = 0.20$$

$$S_1 = R_1 = W_1 = \frac{1 - 0.20}{3} \approx 0.2667$$

$$\text{or} \quad S_1 = R_1 = W_1 = 0.273 - \frac{0.002 \times 0.273}{1 - 0.273} \approx 0.2667$$

The next point can be calculated on the basis of its previous point using the same formulas. These calculations can be performed easily in a worksheet of Ms-Excel®. The columns A, B, C, D, E, and F are reserved for the terms No., M, S, R, W and the response DC-PER, respectively, in the worksheet (Figure 4.9). All points of M should be calculated according to the limits of its varying range, namely, the lower- and the upper-boundary restrictions of 0.18 and 0.30, and

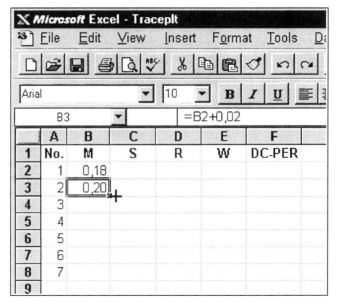

Figure 4.9 Calculating the first two coordinate values of water M.

an increment of 0.02 M is therefore changed from 0.18 to 0.30 with a step of 0.02 in making its response trace plot.

Type the lower restriction value 0.18 in cell B2, and move the cursor to cell B3 with the ↓ key or with one mouse click when the mouse arrow (↖) is located on this cell. Then cell B3 will become surrounded by a black frame as shown in Figure 4.9. Type the equal mark "=" from the keyboard, write behind it "B2 + 0.02", and press the RETURN key (↵). A value of 0.20 will be shown in the blank cell B3. Move the mouse arrow to the right-bottom corner of cell B3, and the arrow (↖) will change into a black cross (+); hold the left mouse key pressed and draw the mouse arrow from cell B3 to B8. After releasing the left mouse key, values such as 0.22, 0.24, 0.26, 0.28, and 0.30 will be shown from cells B4 to B8, respectively. In this way, all the points of M to be drawn in the response trace plot are calculated (Figure 4.10).

The next step is to calculate the coordinate values of soybean flour. Move the cursor to cell C2, type the equal symbol and the equation $(1 - \$B2)/3$ behind it, and press the RETURN key (↵). The value of 0.2733 will be shown in cell C2 (Figure 4.11). The dollar symbol $ in front of B2 ensures the changing of M coordinate values in the formula only with the changing of row but not that of the column. Move the mouse cursor to the right-bottom corner of cell C2 and the mouse arrow form will change from ↖ to +. Draw the mouse from cell C2 to C8 while holding the left mouse key pressed. In cells C3 to C8 the numbers of 0.2733 to 0.2333 will be shown (Figure 4.12). Then the corresponding coordinate values of soybean are calculated.

Drag the mouse from the right-bottom corner of cell C8 (the position in Figure 4.12) to cell E8 while holding the left mouse key pressed. In columns

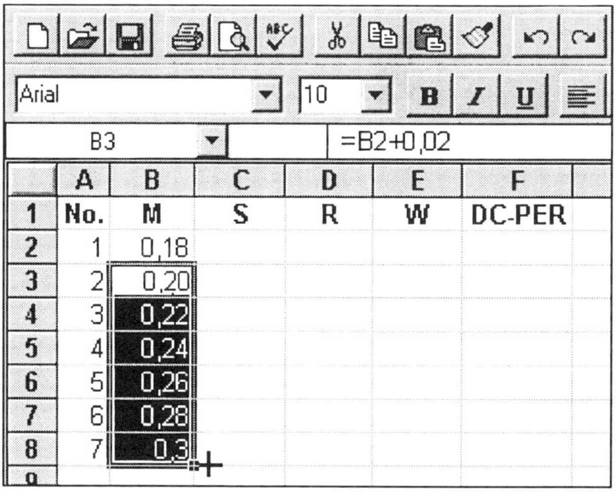

Figure 4.10 Calculating all the coordinate values of water M.

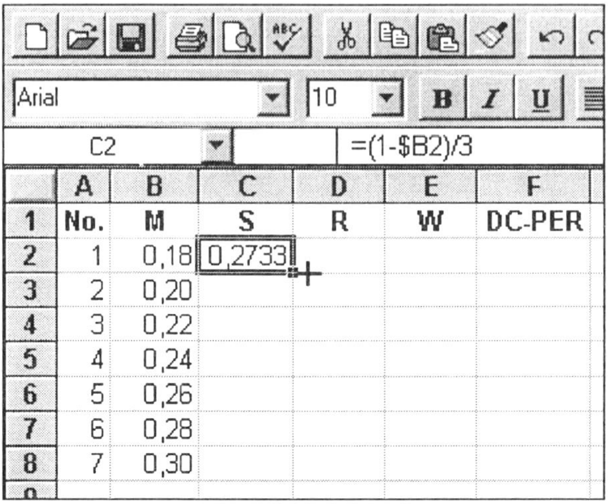

Figure 4.11 Calculating the first coordinate value of soybean S.

D and E the corresponding coordinate values of rice R and wheat W will be calculated out (Figure 4.13).

To calculate the response values (DC-PER), first move the cursor to cell F2. Type the equal symbol and Equation (4.22). Replace the terms M, S, R, and W in the equation with B2, C2, D2, and E2, respectively. Press the RETURN key afterward, and the DC-PER value of 1.837 corresponding to mixture "No. 1" listed in row 2 will appear in cell F2 (Figure 4.14). All the response values corresponding to mixture 2–7 can be calculated by drawing the mouse arrow from cell F2 to F8.

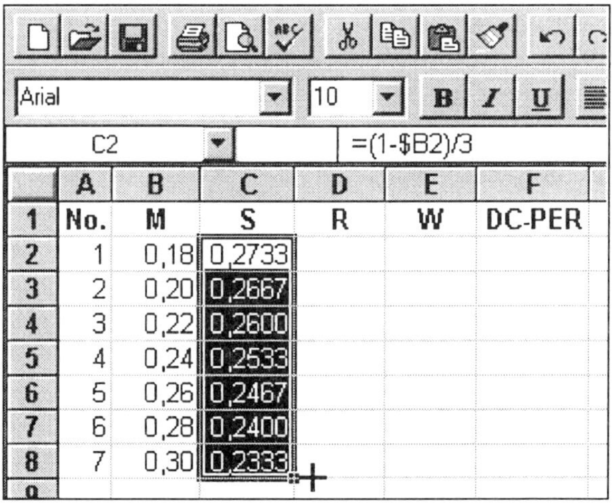

Figure 4.12 Calculating the coordinate values of soybean S.

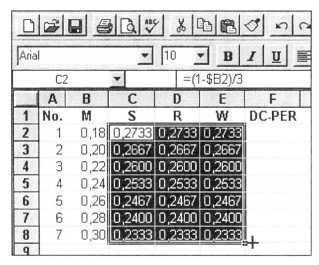

Figure 4.13 Calculating all the coordinate values of rice and wheat.

The next step is to create the response trace plot. Select columns B1 to B8 with the mouse, and hold the Ctrl key pressed while selecting cell F1 to F8. Release the Ctrl key and the left mouse key, move the mouse arrow to the symbol in the toolbar and click on it. The background of this symbol will turn white, as if it had been really pressed. Move the mouse arrow back to the worksheet field and the mouse arrow will change now into the symbol (Figure 4.15). While holding the left mouse key pressed, drag the mouse arrow to create a rectangle (black dotted line) on the worksheet in which the trace plot should be

	A	B	C	D	E	F
				=1,478*B2+2,079*C2+0,585*D2+0,761*E2+		
				4,022*C2*D2+3,635*C2*E2+0,845*D2* E2		
1	No.	M	S	R	W	DC-PER
2	1	0,18	0,2733	0,2733	0,2733	1,837
3	2	0,20	0,2667	0,2667	0,2667	
4	3	0,22	0,2600	0,2600	0,2600	
5	4	0,24	0,2533	0,2533	0,2533	
6	5	0,26	0,2467	0,2467	0,2467	
7	6	0,28	0,2400	0,2400	0,2400	
8	7	0,30	0,2333	0,2333	0,2333	

Figure 4.14 Calculating the first DC-PER value.

	A	B	C	D	E	F	G
1	No.	M	S	R	W	DC-PER	
2	1	0,18	0,2733	0,2733	0,2733	1,837	
3	2	0,20	0,2667	0,2667	0,2667	1,814	
4	3	0,22	0,2600	0,2600	0,2600	1,790	
5	4	0,24	0,2533	0,2533	0,2533	1,768	
6	5	0,26	0,2467	0,2467	0,2467	1,746	
7	6	0,28	0,2400	0,2400	0,2400	1,726	
8	7	0,30	0,2333	0,2333	0,2333	1,705	
9							

Figure 4.15 Selecting the two columns B and F for plot generating.

generated. After releasing the left mouse key, an interactive window will pop up. Do all further steps by following the instructions in the popped up windows (see Section 3.7.1.1 and Figure 3.15) until the trace plot is created (Figure 4.16). The generated plot could be modified and copied for any further application under Ms-Windows®. Figure 4.17 was produced, for instance, by copying the diagram

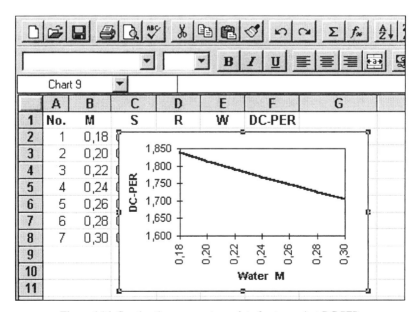

Figure 4.16 Creating the response trace plot of water against DC-PER.

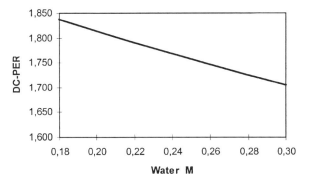

Figure 4.17 Response trace plot of water against DC-PER.

from Ms-Excel® and added into Ms-Word® for Windows®. It is obvious that an increased moisture content leads to extrudates with a decreased protein in vitro nutritional quality. In the same way the response trace plot of soybean, rice, and wheat can be constructed for the analysis of their effects on the DC-PER.

4.7.2 RESPONSE SURFACE PLOT

Similar to the explanation given in Section 3.7.1.2, the effects of mixture components or variables can be illustrated with a response surface assumed to be continuous for all possible mixtures in the simplex region. In practice, a response surface can be a real surface in three-dimensional (3D) space or only a contour plot in 2D surface. Different from the coordinate systems used in Chapter 3, the simplex coordinate system must be adopted for the 3D response or 2D contour plotting. Owing to limited presentation possibilities, a response surface graphic can contain only three variables (components) and one or more responses (food quality indices). If there are more than three components or variables in a mixture model, then only the three relevant variables should be selected for plotting the response surface, whereas the rest of the variables are held at certain levels, normally at their middle or boundary levels. In geometrical terms, each contour curve in a 2D three-component triangle is a projection of a cross section of the 3D surface onto the triangle surface. The whole contour is made by a series of planes parallel to the triangle surface cutting through the surface, usually at a particular height interval.

4.7.2.1 Simplex Coordinate System

To understand the principle of response surface plotting, some basic knowledge about the simplex coordinate system as well as the graphic-generating techniques are presented first. Figure 4.18 demonstrates how a mixture variable works in a simplex coordinate system. In the mixture system three variables

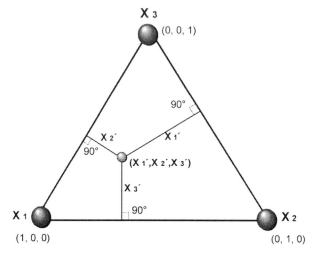

Figure 4.18 Coordinate system of a triangle coordinate system.

or components X_1, X_2, and X_3, which are located at the three vertices of the simplex, are included. The whole mixture system is then enclosed in this simplex. Any mixture, say, (X_1', X_2', X_3'), can be located in this region with the shortest distances to these three simplex edges being X_1', X_2', X_3', respectively. It is not difficult to demonstrate that the sum of X_1', X_2', and X_3' is a definite value, namely 100% or 1. If the levels of any one or two variables of X_1', X_2', and X_3' have been changed, then the level of at least one other variable will vary accordingly. Hence, in a response plot of a three-component system, the response depends in principle only on any two of the three mixture variables. The levels of the third component will be defined automatically.

4.7.2.2 2D Simplex Contour Plot

A contour plot consists of several contour lines or curves in a simplex coordinate system. To draw one contour curve is to determine those mixtures within the experimental region (or region of interest) that lead to a definite value of the response according to the model. These mixtures are then presented geometrically as points in the simplex coordinate system and linked by a curve or a straight line. This whole curve or straight line represents all the mixtures leading to the given response value, which is usually printed on or beside the curve, and must be given in advance so that its contour curve can be drawn.

Unfortunately, there are few computer software programs that can be used to generate a correct 2D contour plot based on a mixture model. Hence, the contour plots may sometimes be drawn by the food researcher himself. In the

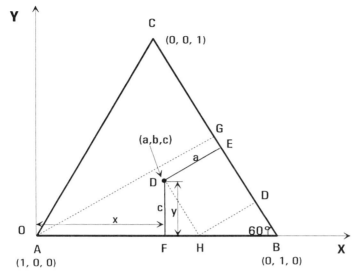

Figure 4.19 Coordinate system of a triangle coordinate system.

following section the basic geometrical principle of a 2D contour, which can be helpful in software programming, is introduced.

4.7.2.2.1 Principle of Coordinate Transformation

On a piece of paper or on the computer screen there are really only two dimensions available. Thus, to locate a point on this surface, only two coordinate values are necessary and possible, e.g., a point can be located in an X-Y coordinate system with its coordinates (X', Y'). To enable a researcher to generate a contour plot in a simplex coordinate system, the simplex coordinates must be transformed into X-Y coordinates so that these curves can be plotted. Figure 4.19 shows the principle of this transformation and the transformation formulas are given in Equations (4.23) and (4.24).

$$AB = \frac{AG}{\sin 60°} = \frac{1}{\sin 60°}$$

$$BH = \frac{HD}{\sin 60°} = \frac{DE}{\sin 60°} = \frac{a}{\sin 60°}$$

$$FH = DF \cdot ctg60° = c \cdot ctg60°$$

$$x = AB - FH - BH = \frac{1}{\sin 60°} - \frac{a}{\sin 60°} - c \cdot ctg60° \tag{4.23}$$

$$y = c \tag{4.24}$$

Some additional restrictions on the y values must be considered to have the contour curves or lines drawn within the triangle. Geometrically, all the y values must be limited under the lines AC and BC. Therefore, there are two different cases to define the restrictions on y:

(1) if $x \leq \frac{0.5}{\sin 60°}$ then

$$y \leq x \cdot tg60°$$

(2) if $x > \frac{0.5}{\sin 60°}$ then

$$y \leq \left(\frac{1}{\sin 60°} - x \right) \cdot tg60° \qquad (4.25)$$

4.7.2.2.2 Selection of Reasonable Response Values

The response values are continuous in the experimental simplex region, but only certain values of them are selected for generating contour curves. It is easy to understand that the response has both maximal and minimal values within the experimental region, and the response values of all contours lie between the response maximum and minimum. To clearly investigate the effects of mixture variables, the contours are generally created with response values that have a certain interval to each other. If it has been decided how many contour curves should be drawn in the simplex region of interest, the response values of each contour line can easily be calculated. For example, to create a contour plot with m curves in a mixture experimental region in which the maximum and minimum of the response are R_{max} and R_{min}, respectively, the step S of the contour lines is computed as follows:

$$S = \frac{R_{max} - R_{min}}{m + 1} \qquad (4.26)$$

The response values for generating each contour curve are:

- contour 1: $R_{min} + 1 \cdot S$
- contour 2: $R_{min} + 2 \cdot S$
- contour 3: $R_{min} + 3 \cdot S$
- contour . . .: . . .
- contour m: $R_{max} - 1 \cdot S$

In general the contour curves of R_{min} or R_{max} cannot be drawn out in the simplex because each is merely a point. Sometimes the step S is selected not evenly so that in the most interesting subregions (for example in the subregions involving maximum or minimum) more contour curves can be drawn. This would help the food researcher to analyze the relationship between response and ingredients more precisely. In this case, steps selected in the region of interest are much smaller than those outside this region.

4.7.2.2.3 Creating a Simplex Contour Plot

According to the principle described above, a BASIC program used to generate simplex contour plots is presented in Table 4.11. The original BASIC code can be rewritten or modified to generate any specific simplex contour plots.

TABLE 4.11. A BASIC Program for Creating Simplex Contours.

```
REM This BASIC program generates a simplex contour plot according to a simplex
    model
10  CLEAR : CLS : SCREEN 0, 0, 0
    T = 0: K1 = 0: K2 = 0: K = 0: T1 = 0

REM set the smoothness of the contour line
    LOCATE 1, 15: COLOR 7, 0: PRINT "+-----------------------------------+"
    LOCATE 2, 15: COLOR 7, 0: PRINT "| Contour Plot for Triangle Coordinate
                                System|"
    LOCATE 3, 15: COLOR 7, 0: PRINT "+-----------------------------------+"
    LOCATE 8, 8: PRINT "(1=Low, 2=Normal, 3=High, 4=UltraHigh)"
    LOCATE 7, 5: PRINT "Please set the SMOOTHNESS of the contour line:";
    COLOR 4, 0: INPUT Fineness: COLOR 7; O
    IF Fineness < 1 OR Fineness > 4 THEN GOTO 10
15  IF Fineness = 1 THEN fine = 20: Stepwide = .1
    IF Fineness = 2 THEN fine = 40: Stepwide = .05
    IF Fineness = 3 THEN fine = 300: Stepwide = .005
    IF Fineness = 4 THEN fine = 1100: Stepwide = .001

    DIM XX(fine), YY(fine), XXX(fine), YYY(fine)

  REM set the contour number
17  LOCATE 10, 5: PRINT "Please set the NUMBER of contour Lines(1-18): ";
    COLOR 4, 0: INPUT ContNo: COLOR 7, 0
    IF ContNo = 0 OR ContNo > 18 THEN GOTO 17

REM Input of coefficients
19  COLOR 7, 0: LOCATE 13, 5: PRINT "Please input the COEFFICIENTS:"
    PRINT"   (From: Y=A1*X1+A2*X2+A3*X3+A12*X1*X2+A13*X1*X3
    +A23*X2*X3)"
    PRINT
    PRINT"   A1 ="; : COLOR 4, 0: INPUT A1: COLOR 7, 0
    PRINT"   A2 ="; : COLOR 4, 0: INPUT A2: COLOR 7, 0
    PRINT"   A3 ="; : COLOR 4, 0: INPUT A3: COLOR 7, 0
    PRINT"   A12 ="; : COLOR 4, 0: INPUT A12: COLOR 7, 0
    PRINT"   A13 ="; : COLOR 4, 0: INPUT A13: COLOR 7, 0
    PRINT"   A23 ="; : COLOR 4, 0: INPUT A23: COLOR 7, 0
    PRINT"  Are the Inputs correct (Y/N) ?";
      REM set a path to change the inputs
21  F$ = INKEY$
    IF F$ ="" THEN GOTO 21
    IF F$ = "N" OR F$ = "n" THEN GOTO 19
    IF F$ = "Y" OR F$ = "y" THEN GOTO 22 ELSE GOTO 21
```

(continued)

TABLE 4.11. (Continued)

```
        REM Check: For quadrate model, A23 shouldn't be 0!!!
22      IF A23 <> 0 THEN GOTO 24
        IF A12 = 0 AND A13 = 0 THEN GOTO 1000

REM Warning in case of A23=0
        BEEP: COLOR 14, 7: LOCATE 24, 23: PRINT"  Attention: Here A23=0 is
        not allowed!   ";
                LOCATE 25, 23: PRINT"  Please exchange X1 with X2 or X3!   ";
23      IF INKEY$ = ""THEN GOTO 23
        REM Blend the field of data input and give warning sign
        COLOR 0, 0: LOCATE 24, 23: PRINT"  Attention: Here A23 = 0 is
        not allowed!   ";
            LOCATE 25, 23: PRINT"  Please exchange the X1 with X2 or X3!   ";
        LOCATE 13, 5: PRINT "Please input the COEFFICIENTS:"
            PRINT"  (Form: Y=A1*X1+A2*X2+A3*X3+A12*X1*X2+A13*X1*X3
            +A23*X2*X3)"
PRINT
        PRINT"  A1 =      ";
        PRINT"  A2 =      ";
        PRINT"  A3 =      ";
        PRINT"  A12 =      ";
        PRINT"  A13 =      ";
        PRINT"  A23 =      ";
        PRINT"  Are the Inputs correct (Y/N) ?   "
GOTO 19

REM Contour data calculation
        REM Find the maximal and minimal values in the region
24      Ymin = A3
        Ymax = A3
        FOR X1 = 0 TO 1.05 STEP .05
        FOR X2 = 0 TO 1.05 STEP .05
        X3 = 1 - X1 - X2
        IF X3 < -.025 THEN GOTO 30
        Y = A1*X1+A2*X2+A3*X3+A12*X1*X2+A13*X1*X3+A23*X2*X3
        IF Y < Ymax THEN GOTO 27
        IF Y > Ymax THEN Ymax = Y: GOTO 30
27      IF Y < Ymin THEN Ymin = Y
30      NEXT X2
        NEXT X1

REM Set the screen to high resolution display mode
        SCREEN 12

        REM Initiation and draw the coordinate: 50-93 to readjust the coordinate
VIEW (50, 93)-(400, 400)
        WINDOW (0, 0)-(1, .866)
        LINE (0, 0)-(.5, .866)
        LINE (0, 0)-(1, 0)
        LINE (.5, .866)-(1, 0)
        LOCATE 25.5, 4: PRINT "X1"
        LOCATE 5, 27.5: PRINT "X2"
        LOCATE 25.5, 52: PRINT "X3"
```

(continued)

158

TABLE 4.11. (Continued)

```
    REM Print out the Model
    LOCATE 28, 5: PRINT "A1/A2/A3/A12/A13/A23:"; A1; A2; A3; A12; A13; A23;

REM Calculate the step according to the contour number
50  Number = (Ymax - Ymin)/ContNo
        REM print minimal and maximal values
        LOCATE 5, 60: PRINT "Ymin ="; Ymin;
        LOCATE 6, 60: PRINT "Ymax ="; Ymax;
        LOCATE 7, 58: PRINT "-----------"

REM Contour smoothness: the smaller the step, the smoother the contour
    FOR Y = Ymin TO Ymax STEP Number
    K1 = 0: K2 = 0
    REM Set the index for contour value to be corrected
    IF T1 = 1 THEN T = T + 1 ELSE T = 1

    FOR X1 = 0 TO 1 STEP Stepwide
    delta = (A2 - A3 + A12 * X1 + A23 - A13 * - X1 - A23 * X1) ^ 2 - 4 * A23
(Y + A3 * X1 + A13 * X1 * X1 - A1 * X1 - A3 - A13 * X1)
    IF delta < 0 THEN GOTO 90: REM no solution

    REM First solution, +
    X21 = ((A2 - A3 + A12 * X1 + A23 - A13 * X1 - A23 * X1) + (delta) ^ .5)/(2 * A23)

    REM Change the coordinate value and check the bound restriction
        K1 = K1 + 1
        XX11 = X1 + X21 * .5
        XX(K1) = 1 - XX11
        YY(K1) = X21 * .866
    IF XX(K1) <= .5 THEN Y21max = .866 + (XX(K1) - .5) * 1.732
                ELSE Y21max = .866 - (XX(K1) - .5) * 1.732
    IF YY(K1) < 0 OR YY(K1) > Y21max THEN GOTO 90
        REM Contour draw
        IF K1 <= 1 THEN 90

COLOR 3: LINE (XX(K1), YY(K1))-(XX(K1 - 1), YY(K1 - 1)
        REM Print the contour response value
        LOCATE T + 8, 60: PRINT "Y("; T; ") ="; Y;
        T1 = 1

    REM Second solution, -
    X22 = ((A2 - A3 + A12 * X1 + A23 - A13 * X1 - A23 * X1) - (delta)^ .5)/(2 * A23)

    REM Triangle coordinate X-Y vs. coordinate system
        K2 = K2 + 1
          XXX11 = X1 + X22 * .5
        XXX(K2) = 1 - XXX11
        YYY(K2) = X22 * .866
    REM check the bound restrictions
        IF XXX(K2) <= .5 THEN Y22max = .866 + (XXX(K2) - .5) * 1.732 ELSE
Y22max = .866 - (XXX(K2) - .5) * 1.732
        IF YYY(K2) < 0 OR YYY(K2) > Y22max THEN GOTO 90
```

(*continued*)

TABLE 4.11. (Continued)

```
REM Second contour draw
        IF K2 <= 1 THEN GOTO 90
        COLOR 3: LINE (XXX(K2), YYY(K2))-(XXX(K2 - 1)), YYY(K2 - 1))
           REM Print the Contour value
           REM LOCATE t + 8, 60: PRINT "Y("; t; ") ="; Y;

  90 NEXT X1
     NEXT Y

  92  F$ = INKEY$
     IF F$ = ""THEN GOTO 92
     BEEP: LOCATE 29, 54: COLOR 5: PRINT "Draw another contour (Y/N)?";
  95  F$ = INKEY$
     IF F$ = ""THEN GOTO 95
     IF F$ = "N" OR F$ = "n" THEN 96
     IF F$ = "Y" OR F$ = "y" THEN GOTO 10 ELSE GOTO 95

96 CLS : SCREEN 0, 0, 0: PLAY "o1 L 20 beegeee"
LOCATE 10, 13: COLOR 7, 0: PRINT      "+-------------------------------+"
     LOCATE 11, 13: COLOR 7, 0: PRINT "| Thank you for your interest! |"
     LOCATE 12, 13: COLOR 7, 0: PRINT "| Dr. Ruguo Hu, Federal Center for Cereal,
                                        Potato |"
     LOCATE 13, 13: COLOR 7, 0: PRINT "| & Lipid Research, D-32756 Detmold,
                                        Tel: 05231 7410 |"
     LOCATE 14, 13: COLOR 7, 0: PRINT "+-------------------------------+"

     END

REM Route for a simple equation Y = A1*X1+A2*X2+A3*X3
     REM Contour data calculation
     REM Find the maximal and minimal values in the region
1000   Ymin = A3
       Ymax = A3
       FOR X1 = 0 TO 1.05 STEP .05
       FOR X2 = 0 TO 1.05 STEP .05
       X3 = 1 - X1 - X2
       IF X3 < -.025 THEN GOTO 1030
       Y = A1 * X1 + A2 * X2 + A3 * X3
       IF Y < Ymax THEN GOTO 1027
       IF Y > Ymax THEN Ymax = Y: GOTO 1030
1027   IF Y < Ymin THEN Ymin = Y
1030   NEXT X2
       NEXT X1

REM Set the screen to high resolution display mode
   SCREEN 12

REM Initiation and draw the coordinate 50-93 to readjust the triangle
       VIEW (50, 93)-(400, 400)
       WINDOW (0, 0)-(1, .866)
       LINE (0, 0)-(5, .866)
       LINE (0, 0)-(1, 0)
       LINE (.5, .866)-(1, 0)
```

(continued)

160

TABLE 4.11. (Continued)

```
        LOCATE 25.5, 4: PRINT "X1"
        LOCATE 5, 27.5: PRINT "X2"
        LOCATE 25.5, 52: PRINT "X3"
        REM Print out the Model
        LOCATE 28, 15: PRINT "A1/A2/A3: "; A1; A2; A3;

        REM Calculate the step for contour number
1050    Number = (Ymax - Ymin) / ContNo
            LOCATE 5, 60: PRINT "Ymin ="; Ymin;
            LOCATE 6, 60: PRINT "Ymax ="; Ymax;
            LOCATE 7, 58: PRINT "-----------"
REM Contour smoothness: the smaller the step, the finer the contour
FOR Y = Ymin TO Ymax STEP Number

        K1 = 0

        REM Set the index for contour value to be corrected
        IF T1 = 1 THEN T = T + 1 ELSE T = 1

        FOR X1 = 0 TO 1 STEP Stepwide
            X21 = (Y - A3 - A1 * X1 + A3 * X1) / (A2 - A3)
        REM Triangle coordinate vs. X-Y coordinate system
            K1 = K1 + 1
            XX11 = X1 + X21 * .5
            XX(K1) = 1 - XX11
        YY(K1) = X21 * .866

        REM Check the bounds restrictions
            IF XX(K1) <= .5 THEN Y21max = .866 + (XX(K1) - .5) * 1.732 ELSE
            Y21max = .866 - (XX(K1) - .5) * 1.732
            IF YY(K1) < 0 OR YY(K1) > Y21max THEN GOTO 1090

        REM Contour draw
            IF K1 <= 1 THEN 1090
            COLOR 4: LINE (XX(K1), YY(K1))-(XX(K1 - 1), YY(K1 - 1))

        REM Print the contour response value
            LOCATE T + 8, 60: PRINT "Y("; T; ") ="; Y;
            T1 = 1

1090    NEXT X1
        NEXT Y

1092    F$ = INKEY$
        IF F$ = ""THEN GOTO 1092
        BEEP: LOCATE 29, 54: COLOR 5: PRINT "Draw another contour(Y/N)?";

1095    F$ = INKEY$
        IF F$ = ""THEN GOTO 1095
        IF F$ = "N" OR F$ = "n" THEN 96
        IF F$ = "T" OR F$ = "y" THEN GOTO 10 ELSE GOTO 1095

END
```

161

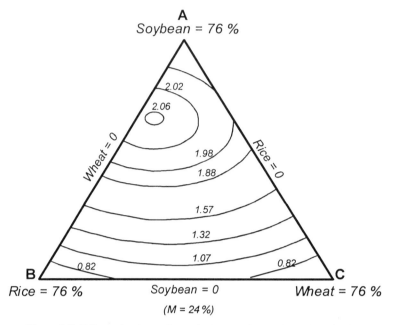

Figure 4.20 Effects of soybean, rice, and wheat on the protein nutritional value.

To create a simplex contour plot according to Model 4.22, the equation must be simplified into Model 4.27 first by fixing the moisture content M at its low level of 24% or any other values within its restrictions. At this moisture content the quality of the extrudates has been found to be quite interesting.

$$DC\text{-}PER = 0.3547 + 2.079 \cdot S + 0.585 \cdot R + 0.761 \cdot W$$
$$+ 4.022 \cdot S \cdot R + 3.635 \cdot S \cdot W + 0.845 \cdot R \cdot W \qquad (4.27)$$

Using a modified version of the program in Table 4.11, a contour plot was generated according to Equation (4.27) and is shown in Figure 4.20. From the numbers beside the curves it can be clearly seen that the normal step of response is 0.25, but in the maximum response region they are set at 0.1 (1.88 to 1.98) and even 0.04 (2.06/2.02/1.98) to grasp the significant changes of the response caused by small changes in the mixture components.

4.7.2.2.4 Reading and Understanding a Simplex Contour Plot

Once a simplex contour graphic is created, you must be able to read and understand it to extract important information about the separate and combined effects of the mixture variables on objective food quality. Unfortunately, a

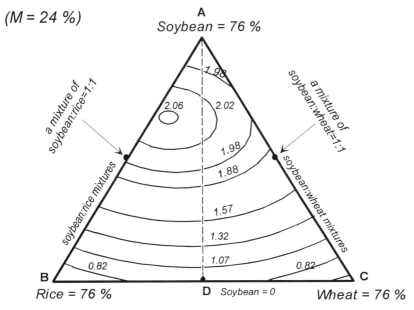

Figure 4.21 Analyzing the effect of soybean in the simplex contour.

simplex contour plot is usually not as easily understood as a normal 2D contour plot. The simplex contour in Figure 4.20 is an example illustrating the effect of soybean on protein nutritional quality (DC-PER) of different extrudates. First a line, real or imaginary, is drawn from the vertex of soybean (A) to the midpoint of the opposite side, BC (dotted line AD in Figure 4.21), on which the percentage of soybean in the recipe is changed from 0% (D, presenting a mixture of rice:wheat:water = 38:38:24) to 76% (A, presenting a mixture of soybean:water = 76:24). In the lower part of the contour, the intervals between two contour curves increase continuously, although the value difference of the DC-PER remains 0.25, namely, from 0.82 to 1.07, 1.07 to 1.32, and 1.32 to 1.57. With the regular and uniform increase of soybean percentage from D to A, it is not difficult to see that a smaller interval between two contour curves means that the response is more sensitive to the percentage change of soybeans in the recipe. Therefore, soybean contributes highly to increasing DC-PER when its levels are lower than approximately 45%. The interval between two contours with the same DC-PER difference within the region of about 76% of soybean (near the top of the plot) is much larger (compared with the smaller contour steps in this region), which means that to increase the soybean percentage in the mixtures will not significantly improve the protein nutritional quality.

The dotted line AD crosses each of the contour curves with the DC-PER of 1.98 and 2.02 twice with the diameter of the latter being smaller than that of the former. This means that only extrudates from formulations with a dominant percentage of soybean, a moderate percentage of rice, and low percentage of wheat can achieve the highest DC-PER value. The highest value of DC-PER in this contour is 2.06, which is printed beside a closed circle. It means that some mixtures can lead to extrudates with a DC-PER value higher than 2.06. In these recipes, a high percentage of soybean but little rice and a small amount of wheat are included.

Generally, subregions in which the highest DC-PER can be achieved are the most important and interesting regions for food product developers. What must be further noticed is that a straight contour line in a simplex triangle does not mean a linear relationship between some variables. This is an essential difference from the contour plot used for process problems (Section 3.7.1.3).

Following the same principles, the effects of rice and wheat on the in vitro protein nutritional quality of the extrudates can be read from their respective contour plots. In addition, the contour shows that the protein quality of a pure rice extrudate (located somewhere lower than contour curve 0.82) is a little better than that of a pure wheat extrudate (notice the additional contour line in the apex C). The DC-PER of pure soybean extrudate is between 1.88 and 1.98. The DC-PER of mixed extrudates decreases rapidly as the percentage of rice or wheat increases in the mixture. The extrudate from a 1:1 mixture of soybean:rice may achieve a better amino acid balance and a higher DC-PER of about 1.99 than that of soybean:wheat of about 1.80. On the other hand, the addition of rice to wheat (line BC) cannot improve the protein quality of the extrudates significantly.

4.7.2.3 3D Simplex Response Surface Graphics

The principles of generating a 3D simplex response surface are quite similar to those discussed in Section 4.7.2.2 for creating 2D simplex contours. In this case, the response should be drawn on the Z-axis in an X-Y-Z 3D simplex coordinate system. Here the surface of X-Y must be projected at a certain angle to achieve a 3D perspective. A typical projection is illustrated in Figure 4.22.

In a 3D simplex coordinate system, only three mixture variables (components) can be involved. If more than three components are included in the mixture model, then only three of them can be drawn in a response graphic simultaneously. Usually, those variables thought to be the most important to the response or of most interest are selected and plotted, so that their effects on the response can be checked graphically.

The general steps of creating a response surface plot for a mixture system with three components X_1, X_2, and X_3 are as follows:

(1) Choose any one variable from the three mixture components, denoted as X_1.

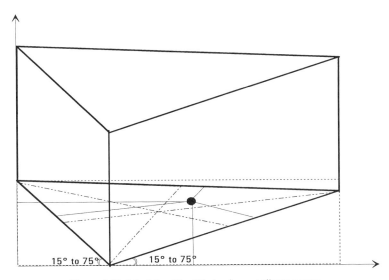

15° to 75°   15° to 75°

Figure 4.22 Principle of the 3D simplex coordinate system.

(2) Fix X_2 at its lowest possible level of X_2^0.

(3) Vary X_1 from its lowest possible value (usually 0 or its lower-bound restriction) to its highest possible value (usually $1 - X_2$ or its upper-bound restriction) with a definite step S_{X1}. If any X_1 value is defined as X_1^i, then $X_1^{i+1} = X_1^i + S_{X1}$. The step is in principle the difference between two nearest values of X_1, which decides the resolution of the response to be generated. Small steps would lead to a diagram with high resolution.

(4) Calculate the corresponding X_3 value X_3^i according to $X_3^i = 1 - X_2^i - X_3^i$ or a similar restriction, such as $X_3^i = a - X_2^i - X_3^i$.

(5) Calculate these series of response values according to the model and these mixtures (X_1^i, X_2^0, X_3^i).

(6) Draw the response on a vertical axis according to its value and coordinates (X_1^i, X_2^0, X_3^i).

(7) Select the varying step S_{X2} for X_2. Fix X_2 on the next smallest value, say $X_2^0 + S_{X2}$.

(8) Repeat steps 3–7.

A 3D simplex response surface graphic can generally supply direct visual information about the effects of the three mixture variables included. The response surface can take many different forms. If the effects among the components are strictly additive, then the surface is depicted by a plane over the simplex. Figure 4.23 shows such a response plane, which represents additive effects of soybean, rice, and wheat on the physiological energy of the extrudates. Obviously, extrudates with higher content of soybean have higher energy levels.

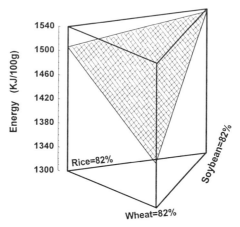

Figure 4.23 Effect of soybean, rice, and wheat on the energy of the extrudates.

4.8 RECIPE OPTIMIZATION AND PREDICTION

Optimization is certainly one of the most important parts of food recipe design. It supplies specific information about which formulations may provide the expected food product quality. Optimization is actually only a purely mathematical calculation, in which the desired food product quality index is defined as the target, and the mixtures or recipe leading to it will be established according to the model. In practice, the optimum formulations are usually not unique depending on a specific objective response. Similar to optimization procedures in a food process system, two basic approaches, numerical and graphical, could be used to perform optimization in a mixture system.

4.8.1 NUMERICAL APPROACHES

4.8.1.1 Optimization with One Objective

In practical food recipe design, the optimum food quality desired is often a maximal or a minimal response value. In this case the optimum mixture(s) can easily be found by calculating the response value of all possible recipes and comparing them to get the optimum response and the corresponding formulation(s). This is the oldest but most reliable way to find an optimum. In the past, such a calculation was almost impossible because laborious computation is required. However, with the capability of modern computers, it is no longer problematic and may be the simplest route for doing optimization. Usually, a fairly small program in any programming language is constructed and used for the optimization calculation. A BASIC program is shown below in Table 4.12 as an example. This program can be modified to meet most optimization objectives in mixture systems.

```
REM A small BASIC program for optimization purposes. There are
   four components a, b, c and d
REM (a+b+c+d=1) in the mixture model. The maximal y value and
   corresponding mixture (a, b, c, d)
REM should be found out. File name "M_optm.bas"
   CLS
OPEN "output.dat" FOR OUTPUT AS #1
PRINT #1, "   ###### Optimization Results ######   "
PRINT #1, "                                        "; date$
PRINT #1
PRINT #1
PRINT #1, "No.:"; " a0"; " b0"; " c0"; " d0"; " y"


let a = 0.5
let b = 0.5
let c = 0.5
let d = 0.5
```

$$y0 = \beta_1{}^*a + \beta_2{}^*b + \beta_3{}^*c + \beta_4{}^*d + \beta_{12}{}^*a^*b + \beta_{13}{}^*a^*c + \beta_{14}{}^*a^*d + \beta_{23}{}^*b^*c + \beta_{24}{}^*b^*d + \beta_{34}{}^*c^*d$$

```
FOR a = 0 TO 1.05 STEP 0.05
   FOR b = 0 TO 1.05 STEP 0.05
      FOR c = 0 TO 1.05 STEP 0.05
            d = 1 - a - b - c
            IF d < 0 THEN 2000
```

$$y = \beta_1{}^*a + \beta_2{}^*b + \beta_3{}^*c + \beta_4{}^*d + \beta_{12}{}^*a^*b + \beta_{13}{}^*a^*c + \beta_{14}{}^*a^*d + \beta_{23}{}^*b^*c + \beta_{24}{}^*b^*d + \beta_{34}{}^*c^*d$$

```
            IF y < y0 THEN GOTO 2000
            i = i + 1
            let y0 = y
            let a0 = a
            let b0 = b
            let c0 = c
            let d0 = d
            PRINT #1, i, ":";a0; ",";b0; ","; c0; ","; d0; ","; y
   2000 NEXT c
   NEXT b
NEXT a

CLOSE #1
PRINT "The optimu is as follow:"
PRINT "a_opt ="; a0
PRINT "b_opt ="; b0
PRINT "c_opt ="; c0
PRINT "d_opt ="; d0
PRINT "Y_opt ="; y0
BEEP
End
```

167

If Model 4.22 is used in this program, then the maximal DC-PER and its corresponding formulation are calculated out as follows:

$$DC\text{-}PER_{max} = 2.18$$
$$M_{opt} = 0.18 \qquad\qquad S_{opt} = 0.58$$
$$R_{opt} = 0.15 \qquad\qquad W_{opt} = 0.09$$

It should be noticed that the optimization can only be within the experimental region for which the model is valid. Any large extrapolation based on the model can generate errors. In the optimization case mentioned above, the varying ranges of water content (M), soybean (S), rice (R), and wheat (W) were 0.18–0.30, 0–0.82, 0–0.82, and 0–0.82, respectively.

4.8.1.2 Optimization with More Than One Objective

Often, the optimum recipe(s) with more than one quality characteristic, described by corresponding models, must be optimized in food product design. In this case, optimization is usually the result of a compromise in which all the optimum responses must be considered simultaneously, with emphasis focused on the more important product quality index. In practice, the sensory quality of a food product is usually taken as the most important target of recipe optimization, especially if the food is an end product intended for consumption after processing. Factors, such as cost, nutritional value, and shelf life, may sometimes also be critical for a recipe optimization. The following is an example of recipe optimization in which two responses were considered at the same time.

A food product with an extremely high sensory evaluation score ($S_{sensory}$) and of a high in vitro calculated protein nutritional quality (DC-PER) would be developed from soybean (S), rice (R), wheat (W), and water (M) through extrusion cooking (extrusion conditions are kept unchanged and assumed to be optimal). As already explained above, the water content has changed from 18% to 30%, therefore, soybean, rice, and wheat can be varied only from 0% to 82%. Two mixture models that describe the sensory evaluation score $S_{sensory}$ [Equation (4.28)] and the DC-PER [Equation (4.22)] would be built based on a simplex centroid design.

$$S_{sensory} = 3.726 \cdot M + 3.135 \cdot S + 3.975 \cdot R + 3.943 \cdot W + 3.336 \cdot S \cdot R \quad (4.28)$$

The optimization task here is to find those recipes that would lead to food extrudates with the highest value of $S_{sensory}$ (most important) and DC-PER. A small program in BASIC can be written to perform the optimization calculation (Table 4.13). It would run under any BASIC environment such as Ms-QBASIC®

TABLE 4.13. A BASIC Program for Optimization of the Sensory and the Nutritional Quality.

```
REM A program to carry out an optimization with two objectives (responses).
CLS
REM Title and date in the optimization file
OPEN "M_optim.dat" FOR OUTPUT AS #1
PRINT #1, "   #######     Optimization Results     #######   "
PRINT #1, "                                "; DATE$
PRINT #1, "                        "
PRINT #1, "                        "
PRINT #1, "                        "
PRINT #1, "No. ", " M ", " S ", " R ", " W ", " Sensory", "DC-PER"
REM Sensory at the centroid point
LET M = .24
LET S = .253
LET R = .253
LET W = .253
Sensory0 = 3.726 * M + 3.135 * S + 3.975 * R + 3.943 * W + 3.336 * S * R

REM   Find out recipes result in extrudates with higher sensory quality than
REM   that of the mixture at the centroid point.
FOR M = .18 TO .30 STEP .05
  FOR S = 0 TO .82 STEP .05
    FOR R = 0 TO .82 STEP .05
          W = 1 - M - S - R
          IF W < 0 THEN 2000
          Sensory = 3.726 * M + 3.135 * S + 3.975 * R + 3.943 * W + 3.336 * S * R
    IF Sensory < Sensory0 THEN GOTO 2000
    PER = 1.478*M + 2.079*S + .585*R + .761*W + 4.022*S*R + 3.635*S*W
        + .845*R*W
                  i = I + 1
                  PRINT #1, i; ":",
                  PRINT #1, USING "##.###"; M;
                  PRINT #1, ",";
                  PRINT #1, USING" ##.###"; S;
                  PRINT #1, ",";
                  PRINT #1, USING" ##.###"; R;
                  PRINT #1, ",";
                  PRINT #1, USING" ##.###"; W;
                  PRINT #1, ",";
                  PRINT #1, USING" ##.###"; Sensory;
                  PRINT #1, USING" ##.###"; PER
2000 NEXT R
    NEXT S
      NEXT M
REM End the program
CLOSE #1
BEEP
END
```

169

found in Ms-DOS® 3.XX or higher versions. In the program, the scanning principle is applied and the $S_{sensory}$ values of all mixtures are calculated step by step. Those recipes leading to extrudates with higher $S_{sensory}$ than that of the mixture at the centroid point ($M:S:R:W = 24:25.3:25.3:25.3$) are recorded with the file names "M_optim.dat". After running the program, one can read this file with a word processor program such as MS-Editor® (included in MS-DOS®), MS-Word®, WordPerfect®, WordStar®, WriterStar®, or AmiPro®. From the file, the highest $S_{sensory}$ value is found easily. Together with the DC-PER data listed, the optimum recipes can be chosen, leading to extrudates with the best sensory quality as well as the highest protein nutritional quality.

The content of file "M_optim.dat" that results from running this program is printed out in Table 4.14. It is obvious that the optimum data are entered in the file "optim.dat" according to the size of the sensory value $S_{sensory}$, namely, from 3.906 to 4.178. A food extrudate with both best sensory quality (4.178) and protein nutritional value (1.703) can be obtained when using the recipe of No. 33. Therefore, No. 33 is the optimum for the given conditions and is shown in Table 4.15.

For a better understanding of optimization, consider the optimization results if the protein nutritional quality (DC-PER) of the extrudates are taken as most important, whereas sensory quality is regarded as the secondary target. Such an optimization can also be performed easily through exchanging the model of DC-PER with that of the sensory score in the program in Table 4.13 and rewriting the judgment "IF Sensory < Sensory0 THEN GOTO 2000" as "IF PER < PER0 THEN GOTO 2000". The content of the file "M_optim.dat" would then be as follows (Table 4.16):

It is clear that the optimization data are entered in the file "M_optim.dat" this time according to the ascending sequence of DC-PER values. The highest DC-PER of 2.178 can be achieved, but with a correspondingly low sensory quality (3.724). If one wants to find an optimum recipe, which would meet part of the requirement for a high sensory quality as well as a high protein nutritional quality, then recipe No. 9 might be the suitable choice. Using this recipe, a food product with a sensory evaluation score of 3.868 and a DC-PER of 1.895 can be extruded, which is a compromise from the previous recipes.

4.8.2 GRAPHICAL APPROACHES

Recipe optimizations can also be performed graphically, or more precisely, be described graphically. If there are only three components in the mixture, then the desired optimum region can be directly read from the response contour plot. Problems would occur if there were more than three mixture components existing in the model because only three can be involved in a triangle graphic. In other words, the rest of the components (variables) must be fixed at certain constant levels. However, these constant levels are not necessarily the optimum

TABLE 4.14. Optimum Recipes of the Sensory Score (with priority) and the DC-PER.

No.	M	S	R	W	####### Optimization Results ####### 12-04-1996 Sensory	DC-PER
1:	0.180,	0.000,	0.050,	0.770,	3.906	0.914
2:	0.180,	0.000,	0.100,	0.720,	3.907	0.933
3:	0.180,	0.000,	0.150,	0.670,	3.909	0.949
4:	0.180,	0.000,	0.200,	0.620,	3.910	0.960
5:	0.180,	0.000,	0.250,	0.570,	3.912	0.966
6:	0.180,	0.000,	0.300,	0.520,	3.914	0.969
7:	0.180,	0.000,	0.350,	0.470,	3.915	0.967
8:	0.180,	0.000,	0.400,	0.420,	3.917	0.962
9:	0.180,	0.000,	0.450,	0.370,	3.918	0.952
10:	0.180,	0.000,	0.500,	0.320,	3.920	0.937
11:	0.180,	0.000,	0.550,	0.270,	3.922	0.919
12:	0.180,	0.000,	0.600,	0.220,	3.923	0.896
13:	0.180,	0.000,	0.650,	0.170,	3.925	0.869
14:	0.180,	0.000,	0.700,	0.120,	3.926	0.838
15:	0.180,	0.000,	0.750,	0.070,	3.928	0.802
16:	0.180,	0.000,	0.800,	0.020,	3.930	0.763
17:	0.180,	0.050,	0.350,	0.420,	3.933	1.165
18:	0.180,	0.050,	0.400,	0.370,	3.943	1.158
19:	0.180,	0.050,	0.450,	0.320,	3.953	1.147
20:	0.180,	0.050,	0.500,	0.270,	3.963	1.132
21:	0.180,	0.050,	0.550,	0.220,	3.973	1.112
22:	0.180,	0.050,	0.600,	0.170,	3.983	1.088
23:	0.180,	0.050,	0.650,	0.120,	3.993	1.060
24:	0.180,	0.050,	0.700,	0.070,	4.003	1.028
25:	0.180,	0.050,	0.750,	0.020,	4.013	0.991
26:	0.180,	0.100,	0.550,	0.170,	4.024	1.287
27:	0.180,	0.100,	0.600,	0.120,	4.043	1.262
28:	0.180,	0.100,	0.650,	0.070,	4.061	1.233
29:	0.180,	0.100,	0.700,	0.020,	4.079	1.199
30:	0.180,	0.150,	0.600,	0.070,	4.102	1.418
31:	0.180,	0.150,	0.650,	0.020,	4.129	1.387
32:	0.180,	0.200,	0.600,	0.020,	4.162	1.555
33:	0.180,	0.250,	0.550,	0.020,	4.178	1.703

TABLE 4.15. Optimum Recipe.

Optimum Recipe				Quality Indices	
M	S	R	W	Sensory	DC-PER
0.180	0.250	0.550	0.020	4.178	1.703

TABLE 4.16. Optimum Recipes of the DC-PER (with priority) and the Sensory Score.

No.	M	S	R	W	####### Optimization Results ####### 12-04-1996 Sensory	DC-PER
1:	0.180,	0.250,	0.100,	0.470,	3.789	1.769
2:	0.180,	0.250,	0.150,	0.420,	3.832	1.779
3:	0.180,	0.250,	0.200,	0.370,	3.875	1.784
4:	0.180,	0.250,	0.250,	0.320,	3.918	1.785
5:	0.180,	0.300,	0.000,	0.520,	3.662	1.853
6:	0.180,	0.300,	0.050,	0.470,	3.713	1.869
7:	0.180,	0.300,	0.100,	0.420,	3.765	1.882
8:	0.180,	0.300,	0.150,	0.370,	3.816	1.890
9:	0.180,	0.300,	0.200,	0.320,	3.868	1.895
10:	0.180,	0.350,	0.000,	0.470,	3.621	1.949
11:	0.180,	0.350,	0.050,	0.420,	3.681	1.965
12:	0.180,	0.350,	0.100,	0.370,	3.741	1.977
13:	0.180,	0.350,	0.150,	0.320,	3.801	1.984
14:	0.180,	0.350,	0.200,	0.270,	3.861	1.987
15:	0.180,	0.400,	0.000,	0.420,	3.581	2.028
16:	0.180,	0.400,	0.050,	0.370,	3.649	2.043
17:	0.180,	0.400,	0.100,	0.320,	3.717	2.053
18:	0.180,	0.400,	0.150,	0.270,	3.786	2.059
19:	0.180,	0.400,	0.200,	0.220,	3.854	2.061
20:	0.180,	0.450,	0.000,	0.370,	3.540	2.088
21:	0.180,	0.450,	0.050,	0.320,	3.617	2.102
22:	0.180,	0.450,	0.100,	0.270,	3.694	2.111
23:	0.180,	0.450,	0.150,	0.220,	3.770	2.116
24:	0.180,	0.450,	0.200,	0.170,	3.847	2.117
25:	0.180,	0.500,	0.000,	0.320,	3.500	2.131
26:	0.180,	0.500,	0.050,	0.270,	3.585	2.143
27:	0.180,	0.500,	0.100,	0.220,	3.670	2.151
28:	0.180,	0.500,	0.150,	0.170,	3.755	2.155
29:	0.180,	0.550,	0.050,	0.220,	3.553	2.166
30:	0.180,	0.550,	0.100,	0.170,	3.646	2.173
31:	0.180,	0.550,	0.150,	0.120,	3.740	2.175
32:	0.180,	0.600,	0.100,	0.120,	3.622	2.176
33:	0.180,	0.600,	0.150,	0.070,	3.724	2.178

ones, unless by accidence or if they are previously selected and checked. So, if more than three mixture components are included in the model, the numerical optimization approach may be the only proper way of finding the optimum recipe(s).

In a mixture system with more than three components, the optimum levels of each component in the optimum formulation can be found by using the numerical optimization method. Choose three variables of interest for plotting, whereas replace the other variables in the model with their optimum levels. This will simplify the model to an equation with only three mixture variables. This

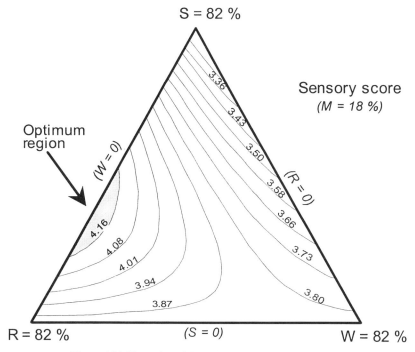

S = 82 %

Sensory score
(M = 18 %)

Optimum
region

(W = 0)

(R = 0)

3.36

3.43

3.50

3.58

3.66

3.73

3.80

4.16

4.08

4.01

3.94

3.87

R = 82 %   (S = 0)   W = 82 %

Figure 4.24 Illustration of the optimum region of sensory score.

simplified model is then used to draw the contours in which the real optimization results will be illustrated.

Let us look at an example of how to illustrate the optimum sensory score of extrudates from soybean, rice, wheat, and water. As shown in Table 4.15, the maximal sensory score (4.183) was achieved at a moisture content of 18%. Thus, for generating the optimum contour plot of the sensory score, the moisture content in the mixtures is fixed at 18%, and a simplified model [Equation (4.29)] is built. This model is used to create a simplex contour plot with the help of a general graphing procedure (Figure 4.24). It can be seen clearly from the contour that the optimum (marked region) is reached with formulations containing a large amount of rice, a moderate amount of soybean, little wheat, and, of course, an 18% water content. The contribution of every component to the product sensory quality is also easily read from this contour. The contour demonstrates the optimum region much more clearly than do the results of the numerical approach.

$$\text{Sensory} = 3.726 \cdot 0.18 + 3.135 \cdot S + 3.975 \cdot R + 3.943 \cdot W + 3.336 \cdot S \cdot R$$

$$= 0.671 + 3.135 \cdot S + 3.975 \cdot R + 3.943 \cdot W + 3.336 \cdot S \cdot R \quad (4.29)$$

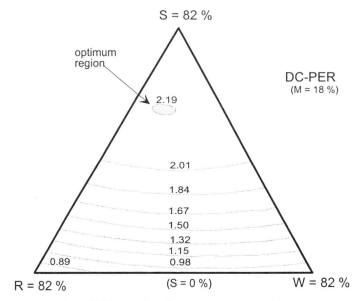

Figure 4.25 Illustration of the optimum region of DC-PER.

Based on the same principle, a contour plot of protein nutritional quality can be generated (Figure 4.25) where the model of DC-PER [see Equation (4.22)] is used and the moisture content M is fixed at 18% (see Table 4.16). It is easy to see from the contour that products with high protein nutritional value (within the filled circle) can be extruded from formulations with more soybean, less rice and wheat, and a water content of 18%.

If the optimization of both sensory quality and protein nutritional value should be considered, then Figures 4.24 and 4.25 can be overlaid to show common optimum regions (Figure 4.26). According to the overlaid graphics, different optimum regions can be identified. For the instance given, two typical optimum regions might be:

- formulations for extruded foods with a sensory score higher than 4.16 and DC-PER more than 2.01
- formulations for food extrudates with DC-PER higher than 2.19 and a sensory score better than 3.80

4.8.3 PREDICTIONS

Prediction is used to estimate the product quality index of some formulations, which are not included in the mixture experimental design and thus not actually tested, by using the mixture model built. The model describes the response surface over the entire simplex space, so that a prediction of the response to any

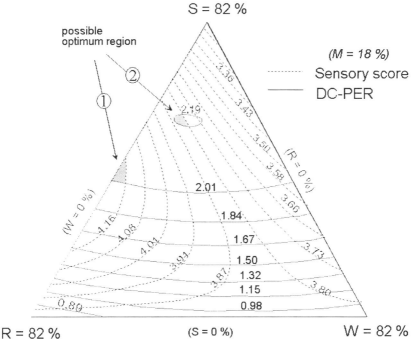

Figure 4.26 Graphical optimization: overlay of contours for sensory quality and DC-PER.

mixture over the entire simplex is possible. Prediction supplies an approach to extract important information from the model. However, the prediction always has a limited statistical accuracy, and in practice there are rarely predictions that are completely correct. Optimization can be considered a specific kind of prediction. Similar to the optimization approach, there are also graphical or mathematical approaches that can be used in performing predictions.

In the numerical approach, all possible formulations can be calculated with a given response value or vice versa. It is usually a simple and purely mathematical calculation. The graphical approach, on the other hand, is based on the simplex contour plot and provides visual and clear overall views of the prediction. According to the response values normally printed on the contour lines, the corresponding formulations can easily be established. Sometimes, if the given response value is not included in the contour lines in the graphic, imaginary interpolation must be used to make the prediction.

CHAPTER 5

Modeling and Optimization for Both Food Recipe and Process

5.1 INTRODUCTION

IN food product development there are often situations in which the process variables and the recipes must be considered and optimized simultaneously. In Chapter 4 the single and joint effects of soybean, rice, wheat, and water on protein nutritional quality and sensory quality of extrudates were studied. All the trials were performed with the same cooking extruder under the same operation conditions. However, in practical food product design, one might also want to study the effect of extrusion conditions, such as mass temperature, screw rotation speed, feed rate, and so on. These operating parameters do not form any part of the formulation or mixture and, therefore, are not component variables; instead, they are process variables.

To solve such a problem, statistical designs in which process and mixture variables are involved must be used to model and optimize the total food system. The usual approach is to combine a simplex centroid design with a complete factorial arrangement as discussed in this chapter.

5.2 PRINCIPLE OF COMBINED DESIGNS

Usually, the test levels of each process variable are limited to two, which can be easily coded as -1 and $+1$. The effects of the n process variables are then examined with a 2^n factorial design; therefore, their region of interest is an n-dimensional hypercube. The simplex centroid mixture design for q components has its own region of interest in a $(q - 1)$-dimensional simplex. The task of statistical experimental planning is to combine these two regions of interest to form a single region for study. This can be achieved through the combination of the factorial experimental plan with the researched mixtures in the simplex centroid design. In other words, to set up the trial plan is to position

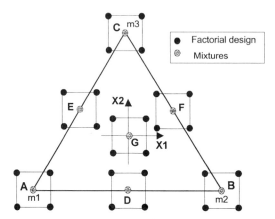

Figure 5.1 The combined design as a result of a 2^2 factorial arrangement set up at each mixture in a three-component simplex centroid design.

a factorial arrangement in the settings of the process variables at each point of composition (formulation) in the mixture components. Therefore, the factorial trials are performed with every formulation in the mixture design.

Figure 5.1 shows an example of a combined experimental plan for two process variables X_1, X_2 and three components $m_1(A)$, $m_2(B)$, and $m_3(C)$. The general trial plan is generated through the combination of the three-component simplex centroid design with $(2^3 - 1)$ mixtures and a 2^2 factorial arrangement with 2^2 trials. Thus, it involves $28 (=(2^3 - 1) \cdot 2^2)$ single trials, and each of the

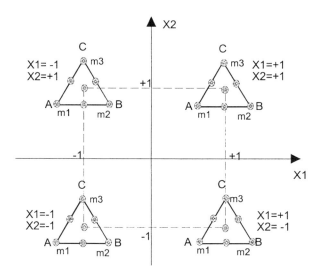

Figure 5.2 The combined design obtained by setting up a three-component mixture simplex centroid design at each 2^2 factorial arrangement.

TABLE 5.1. A Combined Experimental Design for Two Process and Three Mixture Variables Corresponding to Figure 5.1.

		Mixtures			Process				Mixtures			Process	
No.	Code	m_1	m_2	m_3	X_1	X_2	No.	Code	m_1	m_2	m_3	X_1	X_2
1	A	1	0	0	−1	−1	15	D	0.5	0.5	0	+1	−1
2	A	1	0	0	−1	+1	16	C	0.5	0.5	0	+1	+1
3	A	1	0	0	+1	−1	17	E	0.5	0	0.5	−1	−1
4	A	1	0	0	+1	+1	18	E	0.5	0	0.5	−1	+1
5	B	0	1	0	−1	−1	19	E	0.5	0	0.5	+1	−1
6	B	0	1	0	−1	+1	20	E	0.5	0	0.5	+1	+1
7	B	0	1	0	+1	−1	21	F	0	0.5	0.5	−1	−1
8	B	0	1	0	+1	+1	22	F	0	0.5	0.5	−1	+1
9	C	0	0	1	−1	−1	23	F	0	0.5	0.5	+1	−1
10	C	0	0	1	−1	+1	24	F	0	0.5	0.5	+1	+1
11	C	0	0	1	+1	−1	25	G	0.33	0.33	0.33	−1	−1
12	C	0	0	1	+1	+1	26	G	0.33	0.33	0.33	−1	+1
13	D	0.5	0.5	0	−1	−1	27	G	0.33	0.33	0.33	+1	−1
14	D	0.5	0.5	0	−1	+1	28	G	0.33	0.33	0.33	+1	+1

seven points of the simplex centroid design is tested with a complete 2^2 factorial plan. Figure 5.1 can be easily redrawn as Figure 5.2, in which the seven blends of the simplex centroid design are set up at each of the 2^2 points of the factorial arrangement. These two figures actually illustrate the same combined experimental plan and are identical to each other. In Tables 5.1 and 5.2 the experimental plans corresponding to Figures 5.1 and 5.2 are listed.

TABLE 5.2. A Combined Experimental Design for Two Process and Three Mixture Variables Corresponding to Figure 5.2.

		Mixtures			Process				Mixtures			Process	
No.	Code	m_1	m_2	m_3	X_1	X_2	No.	Code	m_1	m_2	m_3	X_1	X_2
1	A	1	0	0	−1	−1	15	A	1	0	0	+1	−1
2	B	0	1	0	−1	−1	16	B	0	1	0	+1	−1
3	C	0	0	1	−1	−1	17	C	0	0	1	+1	−1
4	D	0.5	0.5	0	−1	−1	18	D	0.5	0.5	0	+1	−1
5	E	0.5	0	0.5	−1	−1	19	E	0.5	0	0.5	+1	−1
6	F	0	0.5	0.5	−1	−1	20	F	0	0.5	0.5	+1	−1
7	G	0.33	0.33	0.33	−1	−1	21	G	0.33	0.33	0.33	+1	−1
8	A	1	0	0	−1	+1	22	A	1	0	0	+1	+1
9	B	0	1	0	−1	+1	23	B	0	1	0	+1	+1
10	C	0	0	1	−1	+1	24	C	0	0	1	+1	+1
11	D	0.5	0.5	0	−1	+1	25	D	0.5	0.5	0	+1	+1
12	E	0.5	0	0.5	−1	+1	26	E	0.5	0	0.5	+1	+1
13	F	0	0.5	0.5	−1	+1	27	F	0	0.5	0.5	+1	+1
14	G	0.33	0.33	0.33	−1	+1	28	G	0.33	0.33	0.33	+1	+1

The total number of combined experimental plans becomes quite large when q or n is large. Therefore, in practical food research, it is advisable to keep the number of process variables n and the mixture components q smaller than four so that the trial number needed is not too large for realization.

5.3 MODEL BUILDING

5.3.1 MODEL FORM

The combined design in Table 5.1 is used for collecting data to fit a combined model incorporating the mixture components and the process variables. If the three-component (m_1, m_2, and m_3) mixture system is modeled by a special cubic model and the two process variables (X_1 and X_2) by a linear model with the interaction term $X_1 \cdot X_2$, then the form of the combined model corresponding to the trial plan can be summarized in Equation (5.1):

$$
\begin{aligned}
Z = &\sum_{i=1}^{3} B_i^0 m_i + \sum_{j}\sum_{i<j}^{3}\,^{3} B_{ij}^0 m_i m_j + B_{123}^0 m_1 m_2 m_3 \\
&+ \left[\sum_{i=1}^{3} B_i^1 m_i + \sum_{j}\sum_{i<j}^{3}\,^{3} B_{ij}^1 m_i m_j + B_{123}^1 m_1 m_2 m_3 \right] \cdot X_1 \\
&+ \left[\sum_{i=1}^{3} B_i^2 m_i + \sum_{j}\sum_{i<j}^{3}\,^{3} B_{ij}^2 m_i m_j + B_{123}^2 m_1 m_2 m_3 \right] \cdot X_2 \\
&+ \left[\sum_{i=1}^{3} B_i^{12} m_i + \sum_{j}\sum_{i<j}^{3}\,^{3} B_{ij}^{12} m_i m_j + B_{123}^{12} m_1 m_2 m_3 \right] \cdot X_1 \cdot X_2 \quad (5.1)
\end{aligned}
$$

There are a total of 28 terms in Model 5.1. The physical meaning of each term can be explained as follows:

$\sum_{i=1}^{3} B_i^0 m_i$ is the linear blending portion of the model, and B_i^0 is the expected response to component m_i averaged over all combinations of the levels or values of X_1 and X_2.

$\sum\sum_{i<j} B_{ij}^0 m_i m_j + B_{123}^0 m_1 m_2 m_3$ is the nonlinear blending portion of the model and B_{ij}^0 is a measure of the nonlinear blending between components m_i and m_j averaged over all combinations of the levels of X_1 and X_2.

$[\sum_{i=1}^{3} B_i^1 m_i + \sum\sum_{i<j}^{3} B_{ij}^1 m_i m_j + B_{123}^1 m_1 m_2 m_3] \cdot X_1$ and $[\sum_{i=1}^{3} B_i^2 m_i + \sum\sum_{i<j}^{3} B_{ij}^2 m_i m_j + B_{123}^2 m_1 m_2 m_3] \cdot X_2$ represents the effect of changing the level of process variables X_1 and X_2 on the linear and nonlinear blending properties of the components.

TABLE 5.3. Presentation of Model 5.1 in Table Form.

Term	m_1	m_2	m_3	$m_1\,m_2$	$m_1\,m_3$	$m_2\,m_3$	$m_1\,m_2\,m_3$
Constant	B_1^0	B_2^0	B_3^0	B_{12}^0	B_{13}^0	B_{23}^0	B_{123}^0
X_1	B_1^1	B_2^1	B_3^1	B_{12}^1	B_{13}^1	B_{23}^1	B_{123}^1
X_2	B_1^2	B_2^2	B_3^2	B_{12}^2	B_{13}^2	B_{23}^2	B_{123}^2
$X_1\,X_2$	B_1^{12}	B_2^{12}	B_3^{12}	B_{12}^{12}	B_{13}^{12}	B_{23}^{12}	B_{123}^{12}

$[\sum_{i=1}^{3} B_i^{12} m_i + \sum \sum_{i<j}^{3} B_{ij}^{12} m_i m_j + B_{123}^{12} m_1 m_2 m_3] \cdot X_1 \cdot X_2$ represents the interaction effect of the two process variables on the linear and nonlinear blending properties of the three components.

The model form is quite complicated. It will become clear and easier to comprehend if it is presented and viewed in table form. Usually, the mixture variables are set in a row, and the process variables are set in a column. The coefficients in the model are then filled in on the table. A term in the model can be built through multiplying the values with the corresponding mixture variable in the row and the process variable in the column. For example, Equation (5.1) can be rewritten in another form, as shown in Table 5.3. In the table, each term and the corresponding coefficients are easy to read.

5.3.2 CALCULATION OF COEFFICIENTS

To calculate the estimates of the coefficients in Model 5.1 (Table 5.3), experiments must be performed in advance according to the design in Table 5.2. As described in Chapter 4, four mixture models can then be established under four different process conditions, namely $(X_1, X_2) = (-1, -1), (-1, +1), (+1, -1)$, and $(+1, +1)$:

At $(-1, -1)$: $Z = B_1^{(-1,-1)} \times m_1 + B_2^{(-1,-1)} \times m_2 + B_3^{(-1,-1)} \times m_3$

$$+ B_{12}^{(-1,-1)} \times m_1 \times m_2 + B_{13}^{(-1,-1)} \times m_1 \times m_3 + B_{23}^{(-1,-1)}$$

$$\times m_2 \times m_3 + B_{123}^{(-1,-1)} \times m_2 \times m_2 \times m_3 \qquad (5.2)$$

At $(-1, +1)$: $Z = B_1^{(-1,+1)} \times m_1 + B_2^{(-1,+1)} \times m_2 + B_3^{(-1,+1)} \times m_3$

$$+ B_{12}^{(-1,+1)} \times m_1 \times m_2 + B_{13}^{(-1,+1)} \times m_1 \times m_3 + B_{23}^{(-1,+1)}$$

$$\times m_2 \times m_3 + B_{123}^{(-1,+1)} \times m_2 \times m_2 \times m_3 \qquad (5.3)$$

At $(+1, -1)$: $Z = B_1^{(+1,-1)} \times m_1 + B_2^{(+1,-1)} \times m_2 + B_3^{(+1,-1)} \times m_3$

$$+ B_{12}^{(+1,-1)} \times m_1 \times m_2 + B_{13}^{(+1,-1)} \times m_1 \times m_3 + B_{23}^{(+1,-1)}$$

$$\times m_2 \times m_3 + B_{123}^{(+1,-1)} \times m_2 \times m_2 \times m_3 \qquad (5.4)$$

At $(+1, +1)$: $Z = B_1^{(+1.+1)} \times m_1 + B_2^{(+1.+1)} \times m_2 + B_3^{(+1.+1)} \times m_3$

$$+ B_{12}^{(+1.+1)} \times m_1 \times m_2 + B_{13}^{(+1.+1)} \times m_1 \times m_3 + B_{23}^{(+1.+1)}$$

$$\times m_2 \times m_3 + B_{123}^{(+1.+1)} \times m_2 \times m_2 \times m_3 \qquad (5.5)$$

The terms in Table 5.3 can be divided into four parts depending on which process variables are included. Coefficients are computed according to formulas listed below ($i \leq 3$, $j \leq 3$, $i \neq j$):

- coefficients for the terms without process variables (constant):

$$B_i^0 = (\text{sum of the four } B_i)/4$$

$$= \left(B_i^{(-1.-1)} + B_i^{(-1.+1)} + B_i^{(+1.-1)} + B_i^{(+1.+1)} \right)/4 \qquad (5.6)$$

$$B_{ij}^0 = (\text{sum of the four } B_{ij})/4$$

$$= \left(B_{ij}^{(-1.-1)} + B_{ij}^{(-1.+1)} + B_{ij}^{(+1.-1)} + B_{ij}^{(+1.+1)} \right)/4 \qquad (5.7)$$

$$B_{123}^0 = (\text{sum of the four } B_{123})/4$$

$$= \left(B_{123}^{(-1.-1)} + B_{123}^{(-1.+1)} + B_{123}^{(+1.-1)} + B_{123}^{(+1.+1)} \right)/4 \qquad (5.8)$$

- coefficients for terms including X_1:

$$B_i^1 = [(\text{sum of the } B_i \text{ where } X_1 = +1)$$

$$- (\text{sum of the } B_i \text{ where } X_1 = -1)]/4$$

$$= \left(B_i^{(+1.-1)} + B_i^{(+1.+1)} - B_i^{(-1.-1)} - B_i^{(-1.+1)} \right)/4 \qquad (5.9)$$

$$B_{ij}^1 = [(\text{sum of the } B_{ij} \text{ where } X_1 = +1)$$

$$- (\text{sum of the } B_{ij} \text{ where } X_1 = -1)]/4$$

$$= \left(B_{ij}^{(+1.-1)} + B_{ij}^{(+1.+1)} - B_{ij}^{(-1.-1)} - B_{ij}^{(-1.+1)} \right)/4 \qquad (5.10)$$

$$B_{123}^1 = [(\text{sum of the } B_{123} \text{ where } X_1 = +1)$$

$$- (\text{sum of the } B_{123} \text{ where } X_1 = -1)]/4$$

$$= \left(B_{123}^{(+1.-1)} + B_{123}^{(+1.+1)} - B_{123}^{(-1.-1)} - B_{123}^{(-1.+1)} \right)/4 \qquad (5.11)$$

- coefficients for terms including X_2:

$$B_i^2 = [(\text{sum of the } B_i \text{ where } X_2 = +1)$$
$$- (\text{sum of the } B_i \text{ where } X_2 = -1)]/4$$
$$= \left(B_i^{(-1,+1)} + B_i^{(+1,+1)} - B_i^{(-1,-1)} - B_i^{(+1,-1)}\right)/4 \quad (5.12)$$

$$B_{ij}^2 = [(\text{sum of the } B_{ij} \text{ where } X_2 = +1)$$
$$- (\text{sum of the } B_{ij} \text{ where } X_2 = -1)]/4$$
$$= \left(B_{ij}^{(-1,+1)} + B_{ij}^{(+1,+1)} - B_{ij}^{(-1,-1)} - B_{ij}^{(+1,-1)}\right)/4 \quad (5.13)$$

$$B_{123}^2 = [(\text{sum of the } B_{123} \text{ where } X_2 = +1)$$
$$- (\text{sum of the } B_{123} \text{ where } X_2 = -1)]/4$$
$$= \left(B_{123}^{(-1,+1)} + B_{123}^{(+1,+1)} - B_{123}^{(-1,-1)} - B_{123}^{(+1,-1)}\right)/4 \quad (5.14)$$

- coefficients for terms including $X_1 \times X_2$:

$$B_i^{12} = [(\text{sum of the } B_i \text{ where } X_1 \times X_2 = +1)$$
$$- (\text{sum of the } B_i \text{ where } X_1 \times X_2 = -1)]/4$$
$$= \left(B_i^{(-1,-1)} + B_i^{(+1,+1)} - B_i^{(-1,+1)} - B_i^{(+1,-1)}\right)/4 \quad (5.15)$$

$$B_{ij}^{12} = [(\text{sum of the } B_{ij} \text{ where } X_1 \times X_2 = +1)$$
$$- (\text{sum of the } B_{ij} \text{ where } X_1 \times X_2 = -1)]/4$$
$$= \left(B_{ij}^{(-1,-1)} + B_{ij}^{(+1,+1)} - B_{ij}^{(-1,+1)} - B_{ij}^{(+1,-1)}\right)/4 \quad (5.16)$$

$$B_{123}^{12} = [(\text{sum of the } B_{123} \text{ where } X_1 \times X_2 = +1)$$
$$- (\text{sum of the } B_{123} \text{ where } X_1 \times X_2 = -1)]/4$$
$$= \left(B_{123}^{(-1,-1)} + B_{123}^{(+1,+1)} - B_{123}^{(-1,+1)} - B_{123}^{(+1,-1)}\right)/4 \quad (5.17)$$

Manual computation of the coefficients is not complicated but wearisome. However, it can be performed easily with high precision if a spreadsheet program such as MS-Excel® or Lotus 1-2-3® is used. In the last section of this chapter an example is given to show how to do the computation in Ms-Excel®.

5.4 ANALYSIS OF VARIABLE EFFECTS

As shown in Table 5.3, the general model is usually too complicated for the food product developer to extract information about the effects of each variable on the food quality directly from the model. However, single or joint effects of the variables can be analyzed by holding some variables at certain constant levels and using further graphical or numerical approaches as described in Chapters 3 and 4. More information is given in the example presented in Section 5.6. It must be remembered that a mixture variable is essentially different from a process one. They should not be confused with each other, and their effects must be illustrated with different graphic techniques.

5.5 FOOD QUALITY OPTIMIZATION AND PREDICTION

The two most important applications of the general model are food product optimization and food quality prediction. A small program is usually used to perform the optimization calculations. During programming, attention must be paid to the different and specific varying ranges of the mixture and process variables. The prediction approaches discussed in Chapters 3 and 4 are applicable here as well. More information will be presented in the following section of this chapter.

5.6 PRACTICAL EXAMPLE

5.6.1 SELECTION OF VARIABLES AND RESPONSES

In this example, the statistical techniques discussed in this chapter will be used to develop or design food products from soybean, parboiled rice, and wheat through high-temperature, short-time (HTST) extrusion. Rice and wheat are generally considered important cereals, but their protein contents are low and their protein nutritional quality is rather deficient because of an unbalanced amino acid composition. Soybean has a high protein content of great nutritional value and can thus be used to compensate for the lack of cereal protein by improving the amino acid balance of the cereal-based food. However, soybean has an unpleasant taste that must be controlled and minimized in terms of extrusion cooking. In this example, the sensory score S_m, the protein content C_p, the calculated protein in vitro nutritional quality DC-PER, and the physiological energy E (kJ/100 g in dry basis) of the extrudate are selected as objective quality indices (responses). The developed products should be instant-soluble nutritional foods with good sensory properties. Such food can be used as a food basis and modified with flavors, sugar, salt, and so on to manufacture other instant products.

During product development the four raw materials [Water M (%), Soybean S (% dry basis), Rice R (% d.s.), and Wheat W (% d.s.)] were mixed while

TABLE 5.4. Test Levels of Each Variable.

Item / Variable	Mixtures				Extrusion	
	M (%)	R (%)	W (%)	S (%)	T (°C)	n (min^{-1})
Lower level	18	0	0	0	130	120
Upper level	30	82	82	82	170	160
Code	\	\	\	\	X_1	X_2
Coded lower level	\	\	\	\	+1	+1
Coded upper level	\	\	\	\	−1	−1

the two typical extrusion operating parameters [Mass Temperature T (°C) and Screw Rotation Speed n (min^{-1})] were tested. Based on experience and the results of preliminary trials, the water content in the mixture was restricted to between 18% and 30%. The test levels of each variable are listed in Table 5.4.

5.6.2 EXPERIMENTAL PLAN DESIGN

A simplex centroid mixture design for four components (with upper and lower restrictions of one component) and a complete 2^2 factorial design were selected in the study. The multiple combination of the 12 blends (Figure 5.3 and Table 5.5) from the simplex centroid design and the four factorial trials lead to a general experimental plan with 48 (12×4) trials (Table 5.6 and Figure 5.4). The

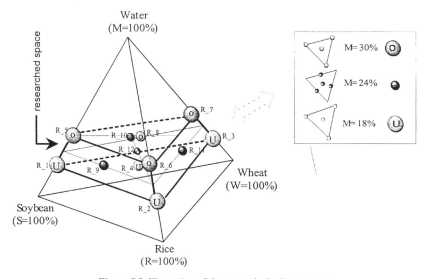

Figure 5.3 Illustration of the researched mixture space.

TABLE 5.5. Researched Formulations.

Recipe No.	M	S	R	W
R_1	0.18	0.82	0	0
R_2	0.18	0	0.82	0
R_3	0.18	0	0	0.82
R_4	0.18	0.273	0.273	0.273
R_5	0.30	0.70	0	0
R_6	0.30	0	0.70	0
R_7	0.30	0	0	0.70
R_8	0.30	0.233	0.233	0.233
R_9	0.24	0.38	0.38	0
R_10	0.24	0.38	0	0.38
R_11	0.24	0	0.38	0.38
R_12	0.24	0.253	0.253	0.253

12 blends build the mixture research region of interest and support the examination and continuous description of the effects of raw materials on the extruded food quality, whereas the 2^2 factorial plan allows the same for the two extrusion variables T and n. The real test levels of T and n are coded into ± 1, and the variables are changed into the coded variables X_1 and X_2 accordingly (Table 5.4).

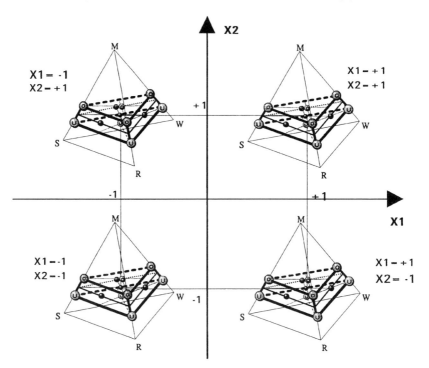

Figure 5.4 Illustration of the general experimental plan.

TABLE 5.6. General Experimental Plan.

No.	Recipe Name	M (×100)	S (×100)	R (×100)	W (×100)	Extrusion X_1	X_2
1	R_1	18	82	0	0	−1	−1
2	R_2	18	0	82	0	−1	−1
3	R_3	18	0	0	82	−1	−1
4	R_4	18	27.3	27.3	27.3	−1	−1
5	R_5	30	70	0	0	−1	−1
6	R_6	30	0	70	0	−1	−1
7	R_7	30	0	0	70	−1	−1
8	R_8	30	23.3	23.3	23.3	−1	−1
9	R_9	24	38	38	0	−1	−1
10	R_10	24	38	0	38	−1	−1
11	R_11	24	0	38	38	−1	−1
12	R_12	24	25.3	25.3	25.3	−1	−1
13	R_1	18	82	0	0	−1	+1
14	R_2	18	0	82	0	−1	+1
15	R_3	18	0	0	82	−1	+1
16	R_4	18	27.3	27.3	27.3	−1	+1
17	R_5	30	70	0	0	−1	+1
18	R_6	30	0	70	0	−1	+1
19	R_7	30	0	0	70	−1	+1
20	R_8	30	23.3	23.3	23.3	−1	+1
21	R_9	24	38	38	0	−1	+1
22	R_10	24	38	0	38	−1	+1
23	R_11	24	0	38	38	−1	+1
24	R_12	24	25.3	25.3	25.3	−1	+1
25	R_1	18	82	0	0	+1	−1
26	R_2	18	0	82	0	+1	−1
27	R_3	18	0	0	82	+1	−1
28	R_4	18	27.3	27.3	27.3	+1	−1
29	R_5	30	70	0	0	+1	−1
30	R_6	30	0	70	0	+1	−1
31	R_7	30	0	0	70	+1	−1
32	R_8	30	23.3	23.3	23.3	+1	−1
33	R_9	24	38	38	0	+1	−1
34	R_10	24	38	0	38	+1	−1
35	R_11	24	0	38	38	+1	−1
36	R_12	24	25.3	25.3	25.3	+1	−1
37	R_1	18	82	0	0	+1	+1
38	R_2	18	0	82	0	+1	+1
39	R_3	18	0	0	82	+1	+1
40	R_4	18	27.3	27.3	27.3	+1	+1
41	R_5	30	70	0	0	+1	+1
42	R_6	30	0	70	0	+1	+1
43	R_7	30	0	0	70	+1	+1
44	R_8	30	23.3	23.3	23.3	+1	+1
45	R_9	24	38	38	0	+1	+1
46	R_10	24	38	0	38	+1	+1
47	R_11	24	0	38	38	+1	+1
48	R_12	24	25.3	25.3	25.3	+1	+1

5.6.3 TRIAL REALIZATION AND RESULTS

The extrudates are dried at room temperature to a moisture content of about 10% and then carefully ground until about 70% of the matter is finer than 200 μm. A 10-gram ground sample is mixed homogeneously with 40 ml of warm water (approximately 40°C) and subjected to sensory evaluation by a panel of assessors. The sensory scores S_m are assessed according to a sensory evaluation scheme similar to that of the German Agriculture Association (DLG; see Appendix) but with slight modifications. The samples were checked in three aspects as follows: Appearance and Color, Texture and Viscosity, and Odor and Taste. The S_m value ranged between 2 and 5; products with evaluating marks higher than 4 were thought to be acceptable.

The protein content C_p (%) is determined according to a quick Kjeldahl method. The conversion factors for soybean, rice, and wheat proteins are 6.25, 5.95, and 5.70, respectively. The C_p of all samples are calculated on a dry weight basis. The DC-PER is used to characterize the protein quality in the study. It is calculated from the amino acid profiles of the proteins in the extrudates with the help of a small program according to the AOAC Official Methods of Analysis No. 43.267. The amino acid composition of the sample protein is determined by using a Chromakon 500 Amino Acid Analyzer. The physiological energy E (kJ/100 g) of the extrudates is calculated on the basis of the recipe and the energy of the raw materials with the assumption that moisture content of all extrudates is adjusted to 9.5%. The experimental results of each food quality index S_m, C_p, DC-PER, and E are listed in Table 5.7.

5.6.4 MODEL BUILDING

Using the responses' data in Table 5.7 and the general experimental plan in Table 5.6, four general models can be built for S_m, C_p, DC-PER, and E (Table 5.8), in which the mixture variables as well as the coded independent extrusion process variables X_1, X_2 are involved [Equations (5.18), (5.19), (5.20), and (5.21)]. In principle they describe systematically and quantitatively the relationships between all the variables and are useful for any further applications such as optimization and prediction of various recipes under different extrusion conditions.

As an example, the model building process of Equation (5.18) (sensory score S_m) is shown step by step below. In the first step, the data, no. 1 to no. 12, no. 13 to no. 24, no. 25 to no. 36, and no. 37 to no. 48 in Table 5.6, are used to build four normal mixture models [Equations (5.22), (5.23), (5.24), and (5.25)] corresponding to extrusion conditions $(X_1, X_2) = (-1, -1), (-1, +1), (+1, -1)$, and $(+1, +1)$:

$$\text{At } (-1, -1): \quad S_m^{(-1, -1)} = 3.80 \cdot M + 2.85 \cdot S + 4.21 \cdot R$$

$$+ 3.59 \cdot W + 3.02 \cdot S \cdot R \qquad (5.22)$$

TABLE 5.7. Experimental Data of S_m, C_p, DC-PER, and E.

No.	Recipe	X_1	X_2	S_m	C_p (%)	DC-PER	E (kJ/100 g)
1	R_1	−1	−1	2.90	39.75	2.301	1540
2	R_2	−1	−1	4.05	9.46	0.783	1511
3	R_3	−1	−1	3.70	11.91	0.924	1378
4	R_4	−1	−1	3.90	19.98	2.080	1477
5	R_5	−1	−1	2.80	39.75	1.971	1540
6	R_6	−1	−1	4.30	9.28	0.675	1511
7	R_7	−1	−1	3.25	11.51	0.885	1378
8	R_8	−1	−1	4.25	20.10	2.067	1477
9	R_9	−1	−1	3.85	24.51	2.072	1526
10	R_10	−1	−1	3.30	25.66	2.066	1459
11	R_11	−1	−1	3.75	10.57	1.057	1444
12	R_12	−1	−1	4.25	20.10	2.066	1477
13	R_1	−1	+1	2.95	39.75	1.970	1540
14	R_2	−1	+1	4.05	9.52	0.872	1511
15	R_3	−1	+1	4.00	11.80	0.961	1378
16	R_4	−1	+1	4.00	19.85	2.067	1477
17	R_5	−1	+1	2.80	39.75	2.076	1540
18	R_6	−1	+1	3.60	9.28	0.733	1511
19	R_7	−1	+1	3.70	11.40	1.025	1378
20	R_8	−1	+1	4.30	20.10	2.066	1477
21	R_9	−1	+1	4.10	23.65	2.068	1526
22	R_10	−1	+1	3.55	25.30	2.068	1459
23	R_11	−1	+1	4.05	10.69	0.707	1444
24	R_12	−1	+1	4.00	19.61	2.068	1477
25	R_1	+1	−1	3.25	40.50	1.971	1540
26	R_2	+1	−1	4.05	9.34	0.486	1511
27	R_3	+1	−1	3.90	11.86	0.608	1378
28	R_4	+1	−1	4.15	20.58	2.203	1477
29	R_5	+1	−1	3.20	39.75	1.969	1540
30	R_6	+1	−1	3.85	9.10	0.557	1511
31	R_7	+1	−1	4.15	11.57	0.657	1378
32	R_8	+1	−1	4.40	20.46	2.067	1477
33	R_9	+1	−1	4.30	24.51	2.066	1526
34	R_10	+1	−1	4.00	25.78	2.066	1459
35	R_11	+1	−1	3.30	10.81	0.813	1444
36	R_12	+1	−1	3.95	20.10	2.085	1477
37	R_1	+1	+1	3.50	40.25	2.078	1540
38	R_2	+1	+1	4.05	9.28	0.450	1511
39	R_3	+1	+1	4.15	11.80	0.588	1378
40	R_4	+1	+1	4.15	20.34	2.176	1477
41	R_5	+1	+1	3.50	39.38	2.079	1540
42	R_6	+1	+1	3.85	9.16	0.572	1511
43	R_7	+1	+1	4.15	11.86	0.821	1378
44	R_8	+1	+1	4.35	20.22	2.066	1477
45	R_9	+1	+1	4.40	24.39	2.066	1526
46	R_10	+1	+1	4.00	25.54	2.066	1459
47	R_11	+1	+1	4.00	10.57	0.827	1444
48	R_12	+1	+1	3.70	20.10	2.077	1477

189

TABLE 5.8. Systematical Models for S_m, C_p, DC-PER, and E.

Term	M	S	R	W	M·S	M·R	M·W	S·R	S·W	R·W
Model for sensory score S_m [Equation (5.18)]										
Constant	3.73	3.14	3.98	3.95	0	0	0	3.34	0	0
X_1	0.41	0.24	−0.12	0.12	0	0	0	−0.11	0	0
X_2	−0.32	0.08	−0.02	0.17	0	0	0	−0.08	0	0
$X_1 X_2$	0.17	0.04	0.10	−0.07	0	0	0	−0.51	0	0
Model for protein content C_p [Equation (5.19)]										
Constant	29.86	39.98	9.38	11.83	9.05	−24.86	−22.80	−0.99	−0.27	0.20
X_1	5.24	0.35	−0.06	0.02	−9.05	−6.34	−4.90	0.99	−0.27	0.20
X_2	7.48	−0.04	0.03	−0.04	−9.05	−8.81	−8.03	−0.99	−0.27	0.20
$X_1 X_2$	−8.04	−0.07	−0.10	−0.02	9.05	9.70	9.86	0.99	−0.27	0.20
Model for protein nutritional quality DC-PER [Equation (5.20)]										
Constant	1.48	2.08	0.59	0.76	0	0	0	4.02	3.64	0.85
X_1	0.24	−0.05	−0.18	−0.19	0	0	0	0.32	0.36	0.85
X_2	0.31	−0.03	−0.03	−0.01	0	0	0	0.01	−0.04	−0.01
$X_1 X_2$	−0.09	0.07	0.01	0.03	0	0	0	−0.12	−0.17	−0.02
Model for physiological energy E [Equation (5.21)]										
Constant	1467	1544	1513	1371	0	0	0	0	0	0
X_1	0	0	0	0	0	0	0	0	0	0
X_2	0	0	0	0	0	0	0	0	0	0
$X_1 X_2$	0	0	0	0	0	0	0	0	0	0

At $(-1, +1)$:
$$S_m^{(-1,+1)} = 2.83 \cdot M + 2.94 \cdot S + 3.97 \cdot R$$
$$+ 4.06 \cdot W + 3.87 \cdot S \cdot R \tag{5.23}$$

At $(+1, -1)$:
$$S_m^{(+1,-1)} = 4.29 \cdot M + 3.26 \cdot S + 3.79 \cdot R$$
$$+ 3.97 \cdot W + 3.81 \cdot S \cdot R \tag{5.24}$$

At $(+1, +1)$:
$$S_m^{(+1,+1)} = 3.99 \cdot M + 3.50 \cdot S + 3.93 \cdot R$$
$$+ 4.16 \cdot W + 2.64 \cdot S \cdot R \tag{5.25}$$

The coefficients of these four models are written into a worksheet of MS-Excel® (Figure 5.5), and the coefficients of the combined model can easily be calculated by using the table calculation function. For computation of the coefficients for terms without extrusion variables, the formula "= (B2 + B3 + B4+B5)/4" was used, and for terms including X_1, X_2 and $X_1 \cdot X_2$, the formulas "= (B4 + B5 − B2 − B3)/4", "= (B3 + B5 − B2 − B4)/4", and "= (B2 + B5−B3−B4)/4" were applied, respectively. These four formulas were typed in corresponding cells B7 to B10. These cells were then selected with the mouse and dragged to column K for calculation of all the coefficients as shown in Figure 5.5.

	A	B	C	D	E	F	G	H	I	J	K
		B7	↓		=(B2+B3+B4+B5)/4						
1		M	S	R	W	M*S	M*R	M*W	S*R	S*W	R*W
2	-1-1	3,80	2,85	4,21	3,59	0,00	0,00	0,00	3,02	0,00	0,00
3	-1+1	2,83	2,94	3,97	4,06	0,00	0,00	0,00	3,87	0,00	0,00
4	+1-1	4,29	3,26	3,79	3,97	0,00	0,00	0,00	3,81	0,00	0,00
5	+1+1	3,99	3,50	3,93	4,16	0,00	0,00	0,00	2,64	0,00	0,00
6											
7	Constant	3,73	3,14	3,98	3,95	0,00	0,00	0,00	3,34	0,00	0,00
8	X1	0,41	0,24	-0,12	0,12	0,00	0,00	0,00	-0,11	0,00	0,00
9	X2	-0,32	0,08	-0,02	0,17	0,00	0,00	0,00	-0,08	0,00	0,00
10	X1*x2	0,17	0,04	0,10	-0,07	0,00	0,00	0,00	-0,51	0,00	0,00
11											

Figure 5.5 Calculating the coefficients in the combined model in a worksheet of Ms-Excel®.

5.6.5 ANALYSIS OF VARIABLE EFFECTS

5.6.5.1 Effects of Raw Materials

The model can be used to graphically analyze the effects of extrusion conditions and raw materials on the four food quality indices. However, too many variables are being included in general Models 5.18, 5.19, and 5.20 to plot their effects. Thus, these three models must be simplified into Equations (5.26), (5.27), and (5.28) by holding mass temperature and screw rotation speed at their middle levels ($X_1 = X_2 = 0$) and the relatively insignificant mixture variable, water content of raw materials, at its middle level of 0.24 (24%):

$$S_m = 0.895 + 3.14 \cdot S + 3.98 \cdot R + 3.95 \cdot W + 3.34 \cdot S \cdot R \qquad (5.26)$$

$$C_p = 7.166 + 39.98 \cdot S + 9.38 \cdot R + 11.83 \cdot W + 2.172 \cdot S - 5.966 \cdot R$$
$$- 5.47 \cdot W - 0.99 \cdot S \cdot R - 0.27 \cdot S \cdot W + 0.20 \cdot R \cdot W \qquad (5.27)$$

$$\text{DC-PER} = 0.355 + 2.08 \cdot S + 0.59 \cdot R + 0.76 \cdot W + 4.02 \cdot S \cdot R$$
$$+ 3.64 \cdot S \cdot W + 0.85 \cdot R \cdot W \qquad (5.28)$$

These three models and Model 5.21 are graphically shown in Figures 5.6, 5.7, 5.8, and 5.9 respectively. From Figure 5.6 it can be concluded that the sensory quality of the extruded products would improve if the proportion of rice and

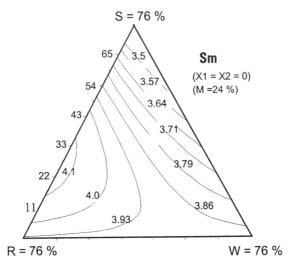

Figure 5.6 Effects of raw materials on the sensory score.

wheat was increased in the formulation, whereas an increased proportion of soybean would lead to food products with a decreased sensory score. Rice contributes more than wheat to the sensory quality. Foods with the highest sensory properties could be produced from mixtures consisting purely of rice and soybean. This means that rice and soybean are better suited to each other with the extrudate taste.

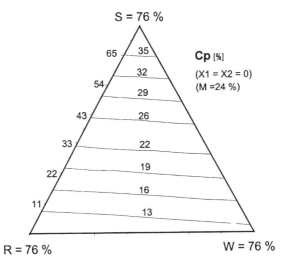

Figure 5.7 Effects of raw materials on the protein content.

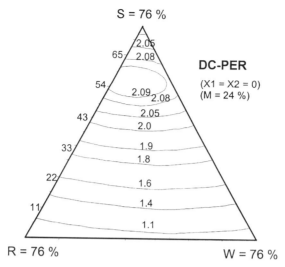

Figure 5.8 Effects of raw materials on the DC-PER.

Figure 5.7 shows that the protein content of the extruded foods increased in almost direct proportion to soybean in the recipe because soybean contains about 40% protein. To the extruded food protein content, wheat contributes more than rice. Figure 5.8 indicates that soybean is the most important raw material affecting the DC-PER. Extruded foods with high protein nutritional

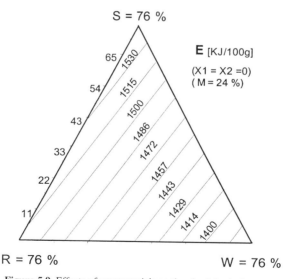

Figure 5.9 Effects of raw materials on the physiological energy.

quality can only be achieved with mixtures consisting of certain amounts of soybean, rice, and wheat.

Figure 5.9 demonstrates that the physiological energy of the extrudates is additive according to the proportions of soybean, rice, and wheat in the formulations. Mixtures containing more soybean would result in food products with higher physiological energy E. However, the difference is significant (up to 20%), because the energy difference between the raw materials is not so large.

5.6.5.2 Effects of Extrusion Cooking

The effects of extrusion cooking on food quality can be illustrated by fixing the recipe at the centroid point (R_12), namely, with "$M\backslash S\backslash R\backslash W = 24\backslash 25.3\backslash 25.3\backslash 25.3$", and simplifying Models 5.18, 5.19, and 5.20 into Equations 5.29, 5.30, and 5.31. The model of physiological energy [Equation (5.21)] does not need to be simplified, for no extrusion variables are included in it. This is also logical because the dry mass of food raw materials (soybean, rice, and wheat) would be lost only insignificantly through extrusion cooking. Thus, the energy of the extruded food would not vary if the energy is characterized on the basis of the dry material substances. Similarly, the protein content in the extruded foods does not change significantly, which can also be concluded from coefficients in the model. Therefore, a graphical analysis of the changes in protein content is not necessary.

$$S_m = 3.90 + 0.15 \cdot X_1 - 0.02 \cdot X_2 + 0.02 \cdot X_1 \cdot X_2 \qquad (5.29)$$

$$C_p = 20.23 + 0.16 \cdot X_1 + 0.14 \cdot X_2 - 0.18 \cdot X_1 \cdot X_2 \qquad (5.30)$$

$$\text{DC-PER} = 1.765 + 0.050 \cdot X_1 + 0.056 \cdot X_2 - 0.015 \cdot X_1 \cdot X_2 \quad (5.31)$$

The graphics in Figures 5.10 and 5.11 are generated according to Models 5.29 and 5.31, respectively. The former graphic shows that the mass temperature affects the sensory score significantly and positively, whereas the screw rotation speed has only a small effect. High temperature and low screw rotation speed results in extruded foods with high sensory properties. The 2D + 3D graphic in Figure 5.11 demonstrates that both the mass temperature and the screw rotation speed affect the DC-PER significantly and with almost the same intensity. With high mass temperature and high screw rotation speed, food products of high protein nutritional quality can be extruded.

5.6.6 OPTIMIZATION OF BOTH RECIPE AND EXTRUSION CONDITION

In principle, various kinds of optimizations can be performed according to the general models 5.18, 5.19, 5.20, and 5.21. A small computer program

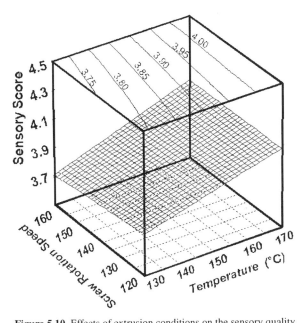

Figure 5.10 Effects of extrusion conditions on the sensory quality.

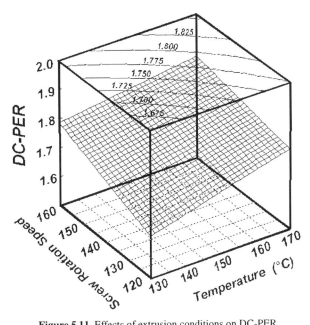

Figure 5.11 Effects of extrusion conditions on DC-PER.

195

TABLE 5.9. Some Results of Optimization.

No. of Optimum Condition	Formulations				Extrusion Conditions			Predicted Food Quality				Criteria for Optimization
	M (%)	S (%)	R (%)	W (%)	T (°C)	n (min^{-1})	S_m	C_p (%)	DC-PER	E (kJ/100 g)	
Opt_1	18	40	40	2	170	120	4.31	24.17	2.01	1516	max. S_m
Opt_2	18	30	50	2	170	120	4.32	20.49	1.80	1513	max. S_m
Opt_3	18	60	10	12	170	160	3.85	31.52	2.27	1508	max. DC-PER
Opt_4	18	60	20	2	170	160	4.00	31.24	2.26	1522	max. DC-PER
Opt_5	19	20	60	1	168	120	4.26	16.89	1.51	1510	S_m & C_p
Opt_6	18	20	50	6	170	120	4.23	17.78	1.55	1501	S_m & C_p

should be written to undertake the calculation during the optimization process. Attention must be paid to the different level ranges of mixture and extrusion variables during programming. Chapters 3 and 4 supplied some examples of programs in BASIC used for the purpose of optimization. They can be modified to perform the optimization here as well.

Some optimum results are shown in Table 5.9. Opt_1 and Opt_2 are the suitable optimum conditions for food products of the highest sensory quality, where S_m is used as the single objective in the optimization process. Likewise, Opt_3 and Opt_4 can be adopted to achieve extrudates with the best protein nutritional quality. If sensory quality and a protein content of about 16% (a value found in most commercial instant children foods) are taken as optimization targets, then Opt_5 and Opt_6 might be the suitable optimum results. In addition, the recipes and extrusion conditions for protein-enriched instant food products with good sensory properties can be predicted as well. Optimizations in which more than two responses are considered as objectives can be realized in a similar way.

The optimization procedure can also be performed graphically. In Figure 5.12, for instance, the optimum regions are shown through the overlay of simplex contours of different responses in the same range of each variable and under the same extrusion conditions. It should be noted that the water content was held at the optimum level of 18% and X_1 and X_2 at their optimum levels for highest S_m and C_p values. With the help of such a graphic, the food developer can generally read the optimum regions visually and understand the location of the optimum regions clearly. For example, black regions in the diagram correspond to such optimum results:

(1) Foods with maximal S_m value (corresponding to Opt_1 & Opt_2)
(2) Foods with relatively high S_m and high DC-PER value (not exactly identical to Opt_3 and Opt_4 because X_2 was not set at its optimum level ($+1$) with respect to the optimization criterion)

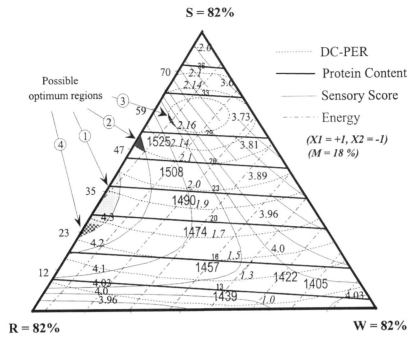

Figure 5.12 Illustration of the optimum regions according to different optimization criteria.

(3) Foods with relatively high DC-PER and high S_m
(4) Foods with high S_m and approximately 17% Cp (⊠, correspond to Opt_5 and Opt_6).

Various kinds of optimizations can be performed or illustrated with such overlaid graphics in a similar way.

CHAPTER 6

Expert System for Food Product Development

6.1 INTRODUCTION

A N expert system is actually a computer program çapable of reasoning and making decisions like a human expert. It is possible to emulate the search behavior of human experts in solving a problem within a special domain of expertise. In an expert system, the subtlety and flexibility of human expertise is combined with the high performance and precision of a computer. Thus, an expert system offers expert advice for making correct and efficient decisions and enables less experienced personnel to perform up to the level of an expert.

Expert systems show a new perspective on information processing, in which the computer serves not only for storage of information or as a calculator but also as an intelligent helper in solving problems such as decision making, process control, and product optimization. With the rapid development of modern computer techniques, the expert system is applied to applications in the field of food product development, food production process control, and automation. The mathematical models built based on statistical experimental methodology can be used as rules in an expert system.

The profitability of a food company, the quality of its food products, and the safety of its operations can be unfavorably affected by some cognitive factors. Having the foresight of exploring and applying expert systems in food product development and production can enhance the chances of successes. In fact, in any food company or manufacturing plant, there are always experts who are usually veteran employees skilled and experienced in some specific knowledge. Such skills come mostly from practical experience accumulated over years, more or less from the technical expertise, and are to some extent difficult to transfer from one individual to another. However, techniques relating to expert systems are well suited to the food processing industry and can assist in improving educational and training efficiency and knowledge transfer.

Of particular interest in the area of food production is the ability of expert systems to be incorporated into the real-time on-line process control, optimization, and fault diagnosis of food processing, to improve the food product quality and the process efficiency. Successful examples of using expert system are seen in atmosphere control of fruit storerooms, control of extruder machinery, control of fermentation reactors, and so on. With the help of an expert system, more time can be devoted to creative output rather than to emergency efforts in case of process failures in a food factory.

The framework of a real-time knowledge-based system is different from that of a rule-based expert system. In a real-time system, the inputs usually come from the automatic sensors that monitor a food process, rather than from operators directly. The outputs of real-time expert systems enable performing a control action on the process, transferring messages to the operators, and possibly recommending desired actions.

6.2 STRUCTURE OF EXPERT SYSTEMS

A simple expert system is composed of a shell and a knowledge base (Figure 6.1). This is the most important characteristic that differentiates expert systems from the conventional computer programs. The shell consists of a user interface, an inference engine (control mechanism), and a data interface that provides the possibility to enter "knowledge" into the knowledge base; in other words, it allows data collection, manual or automatic updating, automatic fault detection, and control actions. The feature of automatic data collection is usually considered as an attractive self-learning capability. It enables the expert system to mimic to some extent the behavior of the human brain.

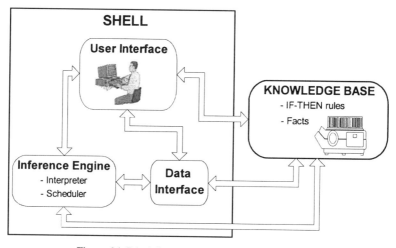

Figure 6.1 Principle structure of a simple expert system.

In the inference engine the general problem-solving knowledge is stored. It usually has access to several different knowledge bases. The inference engine uses information stored in the knowledge base to solve problems. It contains an interpreter that decides how to apply the rules for inferring new knowledge from new rules and facts, and a scheduler that decides when and in what order the rules and facts are applied. Additionally, it records the facts known about the current problem in a database (working memory) that is updated as soon as new information becomes available.

The domain knowledge is collected in the knowledge base. The knowledge base contains facts (data) and rules that are often represented as rule nets. The most widely used expert systems are rule based, in which the expert knowledge is coded in form of "IF-THEN" rules. In this way, the correct solution can be found from all possible solutions. Namely, when the "IF" statement of a rule is fulfilled, the conclusion specified by the "THEN" statement is executed. This often leads to a new fact and further actions that are added to the knowledge base and may match another rule(s) with other facts. In this way all the rules are bound together with each other to form a network (rule net). The mathematical models built based on the statistical experimental methodology can be represented as useful rules.

An expert system is usually designed in such a way that it has a user-friendly interface for easy practical operation. In addition, it should be able to be modified and updated conveniently to improve performance. Expert systems might be the largest success and the most important branch of artificial intelligence. The ability to reason based on knowledge distinguishes the expert systems from conventional computer programs.

6.3 THEORY OF FUZZY LOGIC

In engineering and life sciences, mathematical modeling is commonly applied in terms of classical mathematics and statistics. Exactly formulated mathematical models obtained are based more or less on ideal hypotheses. The relationship between the independent variable and the food quality indices in a concrete practical problem can be described with a mathematical equation. One of the prerequisites for achieving the success of modeling is that all the variable and response data are available as numerical data. However, in practical food production, there are situations in which the food quality and/or independent variables cannot be measured and characterized as numerical data because they are scaled nominally or ordinarily. In addition, some real problems are so complicated and unlimited that a large amount of data for modeling is necessary, which cannot be recognized, treated, or understood by people simultaneously.

Some linguistic food qualities and variables are difficult to interpret into numerical data with traditional methods. For example, if a food product with a sensory score of 4 is thought to be good, then how is the quality of a food with a sensory score of 3.99? Should it be "good" or not? Or in another example, the temperature of a baking oven higher than 250°C can be regarded as very high. Then how about the temperature of 249°C? Is it still "very high" or only "high"? It has been proven that these kinds of concepts cannot be satisfactorily interpreted by the normal exact numerical values.

To solve such problems, the technique of fuzzy logic, introduced by Lotftit A. Zadeh in 1965, is helpful. It provides convenient means to deal with uncertain human judgments and to transfer expert knowledge into quantitative functions that can be processed by a computer. Complex mathematical relationships are normally not required in the construction of fuzzy logic applications.

A fuzzy set is characterized by a set of linguistic values, such as very high, high, normal, low, extra low, and a membership function, which will yield a number between 0 and 1 for any possible linguistic value. In this way, the linguistic statement "the oven temperature is very high" is not just true or false but can be true or false somewhat between 0 and 1. The value "0" denotes exactly "not true at all" or "false", and "1" means "100 percent true." When modeling a process with fuzzy logic technology, both the independent and dependent variables are represented as sets of linguistic values or labels, which in turn are related to numerical values by fuzzy sets. Each linguistic value is represented as a membership function of the value within the interval (0, 1). On the contrary, in the conventional mathematical consideration the grade of membership of each variable gets only values 1 or 0, because it either belongs or does not belong to the fuzzy sets.

Fuzzy logic offers a flexible way to involve uncertainty and subjective expert knowledge in systematic modeling of a food process. The addition of fuzzy logic to an expert system allows the system to focus on the region between qualitative variable values. Fuzzy models are a form of applied artificial intelligence and are based in part on subjective expert knowledge. Fuzzy logic has been popular in Japan and is becoming more and more widespread in Europe and in the United States, especially for products, such as camcorders, cameras, washing machines, vacuum cleaners, and elevators.

6.4 PRINCIPLE OF NEURAL NETWORK

The neural network programs were initially developed as models of human brains, and recently its great potential in the modeling of dynamic, nonlinear systems was realized. It has been widely used in many different fields, such as speech recognition, language processing, character recognition, image feature classification, modeling, and control of different process systems.

Both the fuzzy logic and the neural network are useful and important techniques in the building of a modern expert system in the field of food product design. Just like fuzzy logic, which is characterized by its capability to deal with uncertainties, neural network programs are capable of learning from experiences. They are especially useful when no exact mathematical information on the process under investigation is available. This feature is usually used for both steady-state and dynamic food production control and simulation.

A neural network consists of a great amount of simple processing units (neurons) that are bound to each other. These units can be divided into functional layers, which are organized groups of processing units. A neural network usually has an input layer, one or more hidden layers, and an output layer. The network topology is given by the number of processing units in each layer. Communication with the outside world is one task of the input and output layers. The number of processing units in the input and output layers corresponds to the desired model inputs and outputs, but in the hidden layer the optimal number of the nodes is more or less obtained as a result of trial and error.

The neural network is used to map the inputs to the outputs of the system under investigation. The more complex the relationship between the input and the output data is, more hidden units will be needed. However, too many units can lead to unwanted effects such as learning process noise. Each neuron is usually connected to every other neuron in the next (forward) layer. Every connection between two neurons has a weight factor, which is a real number describing the strength of connection in analogy to the synaptic strength of a neural connection. The inputs, whose values are usually normalized between 0 and 1, are sent parallel to each neuron in the upper layer and multiplied with the weight factors and then summed up. The output of a neuron is obtained from the sum of the normalized weighted values, by evaluating it with a sigmoid transfer function. The same procedure is repeated until the network output is obtained. Different neurons may have different transfer functions. The main function can be summarized as an interpolation from known values to the requested and/or simulated values. In the simplest case, neuron A is linked to another neuron B in the next layer with a certain degree of reliability p, and it can be considered as one rule that "if A is true with degree a then B is true with degree $b = a \cdot p$".

The self-learning capability of the neural network is one of its most important properties. During the learning process, the weights are changed in an iterative way based on the difference between the teacher and the actual output values. The learning method defines how the weights are actually changed. The widely used back-propagation training algorithm is an iterative gradient algorithm designed to minimize the mean square error between the actual output of a multilayer feed-forward network and the desired output.

The features of an expert system depend mainly on knowledge and the way knowledge is represented. The major part of knowledge in an expert system is presented in a network of neurons. A supervised neural network learns from examples through iteration without any requirement of prior knowledge of the relationships of the process variables under investigation.

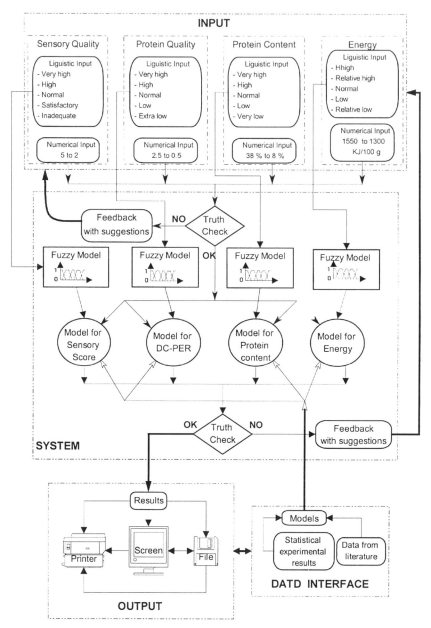

Figure 6.2 Principle sketch of a simple expert system for the development of extruded food product from whole soybean, rice, and wheat.

6.5 BUILDING OF AN EXPERT SYSTEM

Problem identification and goal setting are the first steps toward the creation of an expert system. An obvious bottleneck for the development of all expert systems is the availability of data and knowledge bases. One of the most important steps to set up an expert system is thus to extract and collect the knowledge. For most food systems, the statistical experimental techniques described in this book can be used to reveal the relationship between the food quality indices and the influencing factors in mathematical models. These quantitative rules can be interpreted correspondingly with suitable fuzzy logic sets. The structure of the expert system can then be analyzed and outlined for construction. Usually, each independent variable can be designed as a neuron in the input layer and each of the food quality indices as a neuron in the output layer. The models and additional rules can be designed and integrated in the hidden layer(s). Let us see an example of the application of a simple expert system in the extrusion of instant food as follows.

Cooking extrusion is an important operation unit in many food productions. The goal is to set up an expert system that would give advice on the suitable recipes and extrusion conditions for the development of nutritional, tasteful, instant extruded food products from whole soybean, rice, and wheat flours. Obviously, the nutritional, sensory, and physical qualities of the extruded food are related to the physical properties of the mixtures of incoming ingredients and the operating parameters of the extruder. Even a slight change in operating conditions or raw materials can result in off-grade product. Therefore, statistical experimental methods described in Chapter 5 can be used to build the general regression models for the sensory score, the protein nutritional quality, the protein content and the energy. Fuzzy logic sets can be used to transfer the traditional linguistic description of the product qualities into numerical values for input into the mathematical models obtained. Rules can be established to check the truth of the input and output results of some neurons. The operator will be signaled with tips if these results are found unrealizable.

The principle of the neural network can be used to connect all the units (Figure 6.2). Furthermore, a data interface can be designed to enable the updating of the models used. Even learning algorithms can be involved in the data interface so that the models can be modified automatically based on the results of some specific trials according to the statistical experimental methodology or based on data collected from literature. Of course, such an expert system can also be used for the purpose of food product optimization, prognoses of food quality, process simulation, and so on.

Bibliography

Aguilera, J.M. and F.W. Kosikowski: Soybean extruded product: a response surface analysis. *J. Food Sci.* 41(1976), No. 3, pp. 647–651

Andrews, D.F.: A robust method for multiple linear regression. *Technometrics* 16(1974), No. 4, pp. 523–531

Armstrong, N.A. and C.J. Kenneth: *Pharmaceutical Experimental Design and Interpretation.* Taylor & Francis, 1996

Banks, D.: Is industrial statistics out of control. *Statistical Science* 8(1993), No. 4, pp. 356–409

Barker, T.: *Quality by Experimental Design,* 2nd edition. Marcel Dekker, 1994

Barron, J.J.: Putting fuzzy logic into focus. *Byte* 18(1993), No. 4, pp. 111–118

Bechhofer, R.E., D.M. Goldsman, and T.J. Santner: *Design and Analysis of Experiments for Statistical Selection, Screening, and Multiple Comparisons.* John Wiley & Sons, New York, 1995

Bhat, N. and T.J. Mcavoy: Use of neural nets for dynamic modeling and control of chemical process systems. *Computers Chem. Eng.* 14(1990), No. 4/5, pp. 573–583

Box, G.E.P. and S. Jones: Split-plot designs for robust product experimentation. *Journal of Applied Statistics,* 19(1992), No. 1, pp. 3–26.

Box, G.E.P. and J.S. Hunter: Multi-factor experimental designs for exploring response surfaces. *Annals of Mathematical Statistics* 28(1951), pp. 195–242

Box, G.E.P. and J.S. Hunter: The 2^{k-p} fractional factorial designs. *Technometrics* 3(1961), pp. 311–351

Box, G.E.P. and K.B. Wilson: On the experimental attainment of optimum conditions. *J. Royal Statist. Soc.* 13(1951), No. 1, pp. 1–145.

Box, G.E.P. and N.R. Draper: A basis for the selection of a response surface design. *J. American Statistical Association* 54(1959), pp. 622–654

Box, G.E.P., W.G. Hunter and J.S. Hunter: *Statistics for Experimenters: an Introduction to Design, Data Analysis, and Model Building.* John Wiley & Sons, New York, 1978

Box, G.E.P.: Evolutionary operation: a method for increasing industrial productivity. *Appl. Statist.,* 6(1957), pp. 81–101

Box, G.E.P.: How to Get Lucky. *Quality Engineering,* 5(1993), No. 3, pp. 517–524

Box, G.E.P.: Multi-factor designs of first order. *Biometrika* 39(1952), pp. 49–37

Box, G.E.P.: The exploration and exploitation of response surfaces: some general considerations and examples. *Biometrics* 10(1960), p. 16

Box, G.E.P. and N.R. Draper: *Empirical Model-Building and Response Surface.* John Wiley & Sons, New York, 1987

Brewer, R.F.: *Design of Experiments for Process Improvement and Quality Assurance* (Engineers in Business Series). Institute of Industrial Engineers, 1996

Caudill, M.: Expert networks. *Byte* (1991), No. 10, pp. 108–116

Chitra, S.P.: Use neural networks for problem solving. *Chem. Eng. Progress* 89(1993), No. 4, pp. 44–52

Christensen, L.B.: *Experimental Methodology.* Allyn & Bacon, 1997

Clarke, G.M. and R.E. Kempson: *Introduction to the Design & Analysis of Experiments.* John Wiley & Sons, New York, 1997

Cobb, G.W.: *Introduction to Design and Analysis of Experiments* (Textbooks in Mathematical Sciences). Springer Verlag, 1997

Cochran, W.G. and G.M. Cox: *Experimental Designs.* John Wiley & Sons, New York, 1957

Cook, R.D.: Computer-aided blocking of factorial and response-surface design. *Technometrics* 31(1989), No. 3, pp. 339–346

Cornell, J.A.: Experiments with mixtures: a review. *Technometrics* 15(1973), No. 3, pp. 437– 455

Cornell, J.A.: Some comments on design for Cox's mixture polynomial. *Technometrics* 17(1975), No. 1, pp. 25–35

Cornell, John A. (Ed.): *Experiments with Mixtures: Designs, Models, and the Analysis of Mixture Data.* 2nd edition. John Wiley & Sons, New York, 1990

Davies, O.L.: *The Design and Analysis of Industrial Experiments.* Hafner Publishing Co., New York, 1960

Davies, O.L.: *Statistical Methods in Research and Production.* Oliver and Boyd, Edinburg, 1967

Del Vecchio, R.J.: *Understanding Design of Experiments: A Primer for Technologists* (Hanser Understanding Books). Soho Press, Inc., 1997

Diamond, W.J.: *Practical Experiment Designs for Engineers and Scientists.* 2nd edition. John Wiley & Sons, New York, 1997

Drain, D.: *Handbook of Experimental Methods for Process Improvement.* Chapman & Hall, 1997

Draper, N.R.: An index of rotatability. *Technometrics* 30(1988), No. 1, pp. 105–111

Draper, N.R.: Response surface design for quantitative and qualitative variable. *Technometrics* 30(1988), No. 4, pp. 423–429

Dziezak, J.D.: Taking the grumble out of product development. *Food Technol.* 44(1990), No. 6, pp. 110–117

Eerikäinen, T. and P. Linko: Extrusion cooking modeling, control and optimization. In: *Extrusion Cooking.* Eds.: Mercier, C., P. Linko, and J.M. Harper. American Association of Cereal Chemists, Inc., St. Paul, 1989

Eerikäinen, T., R.J. Aarts, L. Seppä, and P. Linko: Object oriented fuzzy expert system as a tool for extrusion cooker control. In: *Processing and Quality of Foods. Vol. 1, High Temperature/Short Time (HTST) Processing: Guarantee for High Quality Food with Long Shelflife.* Eds.: P. Zeuthen, J.C. Cheftel, C. Eriksson, T.R. Gormley, P. Linko, and K. Paulus. Elsevier Applied Science, London, pp. 1263–1268, 1990

Eilam, M., A. Kinnunen, K.K. Latva, and R. Ahvenainen: Effects of packaging and storage conditions on volatile compounds in gas-packed poultry meat. *Food Additives and Contaminants* 15(1998), No. 2, pp. 217–228

El-Dash, A.A., R. Gonzales, and M. Ciol: Response surface methodology in the control of thermoplastic extrusion of starch. *J. Food Eng.* 2(1983), No. 2, pp. 129–152

Fisher, R.A.: *The Design of Experiments.* Oliver and Boyd, London and Edinburg, 1960

Fishken, D.: Consumer-oriented product optimization. *Food Technol.* 37(1983), No. 11, pp. 49–52

Floros, J.D. and M.S. Chinnan: Computer graphics—assisted optimization for product and process development. *Food Technol.* 42(1988), No. 2, pp. 72–84

Floros, J.D. and M.S. Chinnan: Optimization of pimiento pepper lye-peeling process using response surface methodology. *Trans. ASAE* 30(1987), p. 560

Floros, J.D. and S.C. Manjeet: Computer graphics-assisted optimization. *Food Technol.* (1988), No. 11, pp. 73–84

Floros, J.D.: Computer graphics-assisted optimization for product and process development. *Food Technol.* 42 (1988), No. 2, pp. 72–77

Fowlkes, Y.W. and C.M. Creveling: *Engineering methods for robust product design.* 2nd Edition. Addison-Wesley, Menlo Park, CA, 1997

Frigon, N.L. and D. Mathews: *Practical Guide to Experimental Design.* John Wiley & Sons, New York, 1996

Fukuda, T. and T. Shibata: Theory and applications of neural networks for industrial control systems. *IEEE Trans. Ind. Electron.* 39(1992), pp. 472–489

Gad, S.C.: *Statistics and Experimental Design for Toxicologists.* 3rd edition. CRC Press, Boca Raton, FL, 1998

George Box: *Must We Randomize Our Experiments?* Report Series Report No. 47, University of Wisconsin, December 1989

Giovanni, M.: RSM and product optimization. *Food Technol.* 37(1983), No. 11, pp. 41–45

Giovannitti-Jensen, A.: Graphical assessment of the prediction capability of response surface design. *Technometrics* 31(1989), No. 2, pp. 159–171

Grotch, S.L.: Three-dimensional graphics for scientific data display and analysis. In: *Chemometrics: Mathematics and Statistics in Chemistry.* Eds: B.R. Kowalski, D. Reidel, Dordretch, Holland, 1984

Grove, D.M. and T.P. Davis: *Engineering Quality and Experimental Design.* Longman Scientific and Technical, 1992

Hanneforth, U., W. Seibel, G. Brack, H. Steinhage, H.-G. Ludewig, and R. Haferkamp: Modern Plannungs methoden zur Durchführung von Backversuchen (Systemanalyse, Versuchsplanung, statistische Behandlung and Optimization). *Getreide, Mehl und Brot* 21(1993), No. 1, pp. 47–56

Hare, L.B.: Mixture designs applied to food formulation. *Food Technol.* 28(1974), pp. 50–62

Hartmann, D. and K. Lehner: *Technische Expertensysteme.* Springer-Verlag, Berlin, 1990

Henselman, M.R., S.M. Conatoni, and R.G. Henika: Use of response surface methodology in the development of acceptable high protein bread. *J. Food Sci.* 39(1974), p. 943

Herrod, R.A.: Industrial application of expert systems and the role of the knowledge engineer. *Food Technol.* 43(1989), No. 5, pp. 130–134

Hofer, J.M. and J. Tang: Extrudate temperature control with disturbance predication. *Food Control* 4(1993), pp. 17–24

Hu, R. and K. Seiler: Optimierung der Extrusion von Shrimps-Cracker. *Internationale Zeitschrift für Lebensmitteltechnik, Marketing, Verpackung und Analytik: European Food Science* 44(1993) 10, pp. EFS34–EFS39

Hu, R. and M. Kuhn: Die Anwendung der Central Composite Versuchsplanung (CCD) in der Lebensmitteltechnologie. *Deutsche Lebensmittel-Rundschau* 90(1994), No. 7, pp. 205–211

Hu, R., M. Kuhn, W. Seibel, and K. Seiler: Entwicklung ernährungsphysiologisch und sensorisch optimierter, instantisierter Lebensmittel aus Getreide mit Sojazusatz durch Kochextrusion. *Deutsche Lebensmittel-Rundschau* 92(1996), No. 3, pp. 78–83

Hu, R., M. Kuhn, W. Seibel, and K. Seiler: Modellierung der Veränderungen der Proteinqualität von Getreide und Leguminosen durch Kochextrusion. *Getreide Mehl Brot* 50(1996), No. 6, pp. 373–379

Hu, R., W. Seibel, and K. Seiler: Product Optimization of Food from Starches with Addition of Krill Using Extrusion Cooking and Deep-Frying through System Analytical Method (English). *Proceedings of '94 International Symposium & Exhibition on New Approaches in the Production of Food Stuffs and Intermediate Products from Cereal Grains and Oil Seeds,* Beijing/China, Nov. 16–19, 1994

Hu, R.: Einfluß der Rohstoffe (Soja, Reis, Weizen) und der Kochextrusion auf den Nährwert und die funktionellen Eigenschaften von Extrudaten unter Anwendung statistischer Versuchsmethoden. Ph.D. thesis, Universität Hohenheim, Germany, 1996

Hunter, W.G. and J.R. Kittrell: EVOP, a review. *Technometrics* 8(1966), pp. 389–396

Hunter, W.G. and M.E. Hoff: Planning experiments to increase research efficiency. *Industrial and Engineering Chemistry* 59(1967), pp. 43–48

Jeffrey, T. and V.S. Jordan: *Design of Experiments in Quality Engineering.* McGraw-Hill, New York, 1998

Javanainen, P. and Y. Linko: Factors affecting rye sour dough fermentation with mixed-culture pre-ferment of lactic and propionic acid bacteria. *Journal of Cereal Science* 18(1993), No. 2, pp. 171–185

Jeremiah, P.: Developing a corporate expert systems program. *Food Technol.* 43(1989), No. 5, pp. 155–158

Joglekar, A.M. and A.T. May: Product Excellence through Design of Experiments. *Cereal Foods World* 32(1987), No. 12, pp. 857–868

John, J.A.: *Experiments: Design and Analysis,* 2nd Edition. Charles Griffin & Company Ltd., London and High Wycombe, 1977

Johnson, M.E. and C.J. Nachtsheim: Some guidelines for constructing exact D-Optimal designs on convexdesign spaces. *Technometrics* 25(1983), pp. 271–277

Juran, J.M.: *Juran on Quality by Design.* The Free Press, New York, 1992

Kautola, H.: Itaconic acid production from xylose in repeated-batch and continuous bioreactors. *Applied Microbiology and Biotechnology* 33(1990), No. 1, pp. 7–11

Valentas, Kenneth J., Leon Levine, and J. Peter Clark: *Food Processing Operation and Scale-Up.* Marcel Dekker, Inc., New York, 1991

Kerr, W.: Benefits of statistical experimental designs over single-step experiments. *Cereal Foods World* 41(1996), No. 2, pp. 73–74

Khuri, A.I.: A measure of rotatability for response surface design. *Technometrics* 30(1988), No. 1, pp. 95–104

Kissell, L.T. and B.D. MaTshall: Multi-factor responses of cake quality to basic ingredient ratios. *Cereal Chem.* 39(1962), pp. 16–20

Kotz, S., and N.L. Johnson: *Process Capability Indices.* Chapman and Hall, New York, 1993

Kruse, H., R. Mangold, B. Mechler, and O. Penger: *Programmierung Neuronaler Netze.* Addison-Wesley Publishing Company, Bonn, 1991

Laffey, T.J.: The real-time expert. *Byte* (1991), No. 1, pp. 259–264

Liggett, J.V.: *Dimensional Variation Management Handbook.* Prentice Hall, Englewood Cliffs, NJ, 1993

Linko, P. and Y.H. Zhu: Neural network modeling for real-time variable estimation and predication in the control of glucoamylase fermentation. *Process Biochemistry* 27(1992), pp. 275–283

Linko, P., K. Uemura, and T. Eerikäinen: Neural networks in fuzzy extrusion control. Proceedings of a three-day symposium organized by the Institution of Chemical Engineers' *(IChemE)* Food and Drink Subject Group on behalf of the EFCE Food Working Party. St John's College, Cambridge, 30 March–1 April 1992, pp. 401–410

Logothesis, N. and H.P. Wynn: *Quality through Design.* Clearendon Press, Oxford, 1994

Giovanni, Maria: Response surface methodology and product optimization. *Food Technol.* 1983, Nov., p. 41

McLean, R.A. and V.L. Anderson: Extreme vertices design of mixture experiments. *Technometrics* 8(1966), pp. 447–454

Mclellan, M.R.: Introduction to artificial intelligence and expert system. *Food Technol.* 43(1989), No. 5, pp. 120–128

Mead, R. and D.J. Pike: A review of response surface methodology from a biometric viewpoint. *Biometrics* 31(1975), pp. 803–851

Meuser, F., B. van Lengerich, and F. Köhler: Einfluss der Extrusionsparameter auf funktionelle Eigenschaften von Weizenstärke. *Starch/Stärke* 34(1982), No. 11, pp. 366–372

Microsoft Excel User Manual. Microsoft Cooperation, 1997

Mitchell, T.J.: Computer construction of D-optimal first-order designs. *Technometrics* 20(1974), pp. 211–220

Montagomery, D.C. *Design and Analysis of Experiments.* Georgia Institute of Technology, John Wiley & Sons, New York, 1976

Montgomery, D.C. and R.H. Myers: *Response Surface Methodology: Process and Product in Optimization Using Designed Experiments* (Wiley Series in Probability and Statistics). John Wiley & Sons, New York, 1995

Motycka, R.R., R.E. DeVor, and P.J. Bechtel: Response methodology approach to the optimization of boneless ham yield. *J. Food Sci.* 49(1984), p. 1386

Mustakas, G.C.: New process for low-cost, high-protein beverage base. *Food Technol.* 25(1971), No. 5, pp. 534–541

Myers, R.H.: Response surface methodology: 1966–1988. *Technometrics* 31(1989), No. 2, pp. 137–157

Nachtsheim, C.J.: Tools for computer-aided design of experiments, *J. Quality Technol.* 19(1987), No. 3, p. 132

Peng, K.C.: *The Design and Analysis of Scientific Experiments.* Addison-Wesley Publishing Company, Massachusetts, 1967

Petersen, H.: *Grundlagen der Statistik und der statistischen Versuchsplanung.* Landsberg/Lech: ecomed-Losebl. - Ausg., ecomed Verlagsgesellwchaft GmbH, 1991

Piggott, J.R.: *Statistical Procedures in Food Research.* Elsevier Applied Science, London and New York, 1986

Pinto, J.J.: Application of PC-based expert systems in the processing plant. *Food Technol.* 43(1989), No. 5, pp. 144–1154

Reiner, Wolke: *Fuzzy Logic: Einführung und Leitfaden zur praktischen Anwendung mit Fuzzy-Shell in C++.* Addison-Wesley Publishing Company, Bonn, 1993

Saguy, I. (Ed.): *Computer-Aided Techniques in Food Technology.* Marcel Dekker, Inc., New York, 1983

SAS: SAS/STAT guide. SAS Institute, Cary, NC, 1987

Scheffé, H: Experiments with mixtures. *Journal of the Royal Statistical Society,* 20(1958), pp. 344–360

Scheffé, H.: The simplex-centroid design for experiments with mixtures. *Journal of the Royal Statistical Society,* Series B, 25(1963), pp. 235–263

Schutz, H.G.: Multiple regression approach to optimization. *Food Technol.* (1983), No. 11, pp. 46–48

Schutz, H.G.: Multiple regression approach to optimization. *Food Technol.* 37(1983), No. 11, pp. 46–48

Seibel, W. and R. Hu: Gelatinization characteristics of a cassava/corn starch based blend durch extrusion cooking employing response surface methodology. *Starch/Stärke* 46(1994), No. 6, pp. 217–224

Selwyn, M.R.: *Principles of Experimental Design for the Life Sciences.* CRC Press, Boca Raton, FL, 1996

Shepard, J.D.: Japanese leaders in fuzzy logic. *Byte* 18(1993), No. 4, p. 116

Sherald, M.: Solving the unsolvable. *Byte* (1991), No. 1, pp. 285–288

Shinekey, G.: Expert systems in process control. *Food Technol.* 43(1988), No. 5, pp. 139–142

Shona, D. and A.V.A Holt: Formulation & optimization of tortillas containing wheat, cowpea and peanut flours using mixture response surface methodology. *J. Food Sci.* 57(1992), No. 1, pp. 121–127

Sidel, J.L. and H. Stone: An introduction to optimization research. *Food Technol.* (1983) November, pp. 36–38

Snee, R.D.: Experiments designs for quadratic models in mixture spaces. *Technometrics* 17(1975), No. 2, pp. 149–159

Snee, R.D.: Screening concepts and design for experiments with mixtures. *Technometrics* 18(1976), No. 1, pp. 19–29

Søren Bisgaard: The Early Years of Designed Experiments in Industry: Case Study References and Some Historical Anecdotes. Report Series Report No. 75, University of Wisconsin, November 1991

Søren Bisgaard: The Role of Scientific Problem Solving and Statistics in Quality Improvement: Some Perspectives. European Quality Congress, Trondheim, Norway, June 1997.

SPSS—Statistical Package for the Social Science. McGraw-Hill, New York, 1975

STATISTICA Volume IV. StatSoft Inc., 1997

Wadsworth, H.M.: *Handbook of Statistical Methods for Engineers and Scientists.* 2nd Edition, McGraw-Hill, New York, 1997

Wheeler, A.J. and A.R. Ganji: *Introduction to Engineering Experimentation.* Prentice Hall Press, 1996

Whitney, F.: What expert systems can do for the food Industry. *Food Technol.* 43(1989), No. 5, pp. 135–138

Yandell, B.S.: *Practical Data Analysis for Designed Experiments* (Texts in Statistical Science). Chapman & Hall, 1997

Zadeh, L.A.: Fuzzy sets. *Inform. Control* 8(1965), pp. 338–353

Appendix

A.1 SOME STANDARD CODED EXPERIMENTAL DESIGN

A.1.1 PLACKETT-BURMAN DESIGN

A.1.1.1 Plackett-Burman Design with Twelve Runs (for maximal seven variables): Table 3.3

A.1.1.2 Twenty-Run Plackett-Burman Design (for screening of maximal 15 variables)

No.	Average	X_1	X_2	X_3	X_4	X_5	X_6	X_7	X_8	X_9	X_{10}	X_{11}	X_{12}	X_{13}	X_{14}	X_{15}	X_{16}	X_{17}	X_{18}	X_{19}
1	1	1	1	-1	-1	1	1	1	1	-1	1	-1	1	-1	-1	-1	-1	1	1	-1
2	1	1	-1	-1	1	1	1	-1	1	-1	1	-1	1	-1	-1	-1	1	1	-1	1
3	1	-1	-1	1	1	1	1	-1	1	-1	1	-1	-1	-1	-1	1	1	-1	1	1
4	1	-1	1	1	1	1	-1	1	-1	1	-1	-1	-1	-1	1	1	-1	1	1	-1
5	1	1	1	1	1	-1	1	-1	1	-1	-1	-1	-1	1	1	-1	1	1	-1	-1
6	1	1	1	1	-1	1	-1	1	-1	-1	-1	-1	1	1	-1	1	1	-1	-1	1
7	1	1	1	-1	1	-1	1	-1	-1	-1	-1	1	1	-1	1	1	-1	-1	1	1
8	1	1	-1	1	-1	1	-1	-1	-1	-1	1	1	-1	1	1	-1	-1	1	1	1
9	1	-1	1	-1	1	-1	-1	-1	-1	.1	1	-1	1	1	-1	-1	1	1	1	1
10	1	1	-1	1	-1	-1	-1	-1	1	1	-1	1	1	-1	-1	1	1	1	1	-1
11	1	-1	1	-1	-1	-1	-1	1	1	-1	1	1	-1	-1	1	1	1	1	-1	1
12	1	1	-1	-1	-1	-1	1	1	-1	1	1	-1	-1	1	1	1	1	-1	1	-1
13	1	-1	-1	-1	-1	1	1	-1	1	1	-1	-1	1	1	1	1	-1	1	-1	1
14	1	-1	-1	-1	1	1	-1	1	1	-1	-1	1	1	1	1	-1	1	-1	1	-1
15	1	-1	-1	1	1	-1	1	1	-1	-1	1	1	1	1	-1	1	-1	1	-1	-1
16	1	-1	1	1	-1	1	1	-1	-1	1	1	1	1	-1	1	-1	1	-1	-1	-1
17	1	1	1	-1	1	1	-1	-1	1	1	1	1	-1	1	-1	1	-1	-1	-1	-1
18	1	1	-1	1	1	-1	-1	1	1	1	1	-1	1	-1	1	-1	-1	-1	-1	1
19	1	-1	1	1	-1	-1	1	1	1	1	-1	1	-1	1	-1	-1	-1	-1	1	1
20	1	-1	-1	-1	-1	-1	-1	-1	-1	-1	-1	-1	-1	-1	-1	-1	-1	-1	-1	-1

A.1.2 2^n FACTORIAL DESIGN

A.1.2.1 2^2 Factorial Design

No.	X_1	X_2
1	-1	-1
2	-1	1
3	1	-1
4	1	1

A.1.2.2 2^3 Factorial Design: Table 3.4

A.1.2.3 2^4 Factorial Design: Table 3.7 (first part)

A.1.2.4 2^5 Factorial Design

No.	X_1	X_2	X_3	X_4	X_5	No.	X_1	X_2	X_3	X_4	X_5
1	-1	-1	-1	-1	-1	17	1	-1	-1	-1	-1
2	-1	-1	-1	-1	1	18	1	-1	-1	-1	1
3	-1	-1	-1	1	-1	19	1	-1	-1	1	-1
4	-1	-1	-1	1	1	20	1	-1	-1	1	1
5	-1	-1	1	-1	-1	21	1	-1	1	-1	-1
6	-1	-1	1	-1	1	22	1	-1	1	-1	1
7	-1	-1	1	1	-1	23	1	-1	1	1	-1
8	-1	-1	1	1	1	24	1	-1	1	1	1
9	-1	1	-1	-1	-1	25	1	1	-1	-1	-1
10	-1	1	-1	-1	1	26	1	1	-1	-1	1
11	-1	1	-1	1	-1	27	1	1	-1	1	-1
12	-1	1	-1	1	1	28	1	1	-1	1	1
13	-1	1	1	-1	-1	29	1	1	1	-1	-1
14	-1	1	1	-1	1	30	1	1	1	-1	1
15	-1	1	1	1	-1	31	1	1	1	1	-1
16	-1	1	1	1	1	32	1	1	1	1	1

A.1.2.5 2^6 Factorial Design

No.	X_1	X_2	X_3	X_4	X_5	X_6	No.	X_1	X_2	X_3	X_4	X_5	X_6
1	−1	−1	−1	−1	−1	−1	33	1	−1	−1	−1	−1	−1
2	−1	−1	−1	−1	−1	1	34	1	−1	−1	−1	−1	1
3	−1	−1	−1	−1	1	−1	35	1	−1	−1	−1	1	−1
4	−1	−1	−1	−1	1	1	36	1	−1	−1	−1	1	1
5	−1	−1	−1	1	−1	−1	37	1	−1	−1	1	−1	−1
6	−1	−1	−1	1	−1	1	38	1	−1	−1	1	−1	1
7	−1	−1	−1	1	1	−1	39	1	−1	−1	1	1	−1
8	−1	−1	−1	1	1	1	40	1	−1	−1	1	1	1
9	−1	−1	1	−1	−1	−1	41	1	−1	1	−1	−1	−1
10	−1	−1	1	−1	−1	1	42	1	−1	1	−1	−1	1
11	−1	−1	1	−1	1	−1	43	1	−1	1	−1	1	−1
12	−1	−1	1	−1	1	1	44	1	−1	1	−1	1	1
13	−1	−1	1	1	−1	−1	45	1	−1	1	1	−1	−1
14	−1	−1	1	1	−1	1	46	1	−1	1	1	−1	1
15	−1	−1	1	1	1	−1	47	1	−1	1	1	1	−1
16	−1	−1	1	1	1	1	48	1	−1	1	1	1	1
17	−1	1	−1	−1	−1	−1	49	1	1	−1	−1	−1	−1
18	−1	1	−1	−1	−1	1	50	1	1	−1	−1	−1	1
19	−1	1	−1	−1	1	−1	51	1	1	−1	−1	1	−1
20	−1	1	−1	−1	1	1	52	1	1	−1	−1	1	1
21	−1	1	−1	1	−1	−1	53	1	1	−1	1	−1	−1
22	−1	1	−1	1	−1	1	54	1	1	−1	1	−1	1
23	−1	1	−1	1	1	−1	55	1	1	−1	1	1	−1
24	−1	1	−1	1	1	1	56	1	1	−1	1	1	1
25	−1	1	1	−1	−1	−1	57	1	1	1	−1	−1	−1
26	−1	1	1	−1	−1	1	58	1	1	1	−1	−1	1
27	−1	1	1	−1	1	−1	59	1	1	1	−1	1	−1
28	−1	1	1	−1	1	1	60	1	1	1	−1	1	1
29	−1	1	1	1	−1	−1	61	1	1	1	1	−1	−1
30	−1	1	1	1	−1	1	62	1	1	1	1	−1	1
31	−1	1	1	1	1	−1	63	1	1	1	1	1	−1
32	−1	1	1	1	1	1	64	1	1	1	1	1	1

A.1.3 2^{n-p} FRACTIONAL FACTORIAL DESIGN

A.1.3.1 2^{7-1} Fractional Factorial Design

No.	X_1	X_2	X_3	X_4	X_5	X_6	X_7	No.	X_1	X_2	X_3	X_4	X_5	X_6	X_7
1	−1	−1	−1	−1	−1	−1	1	33	1	−1	−1	−1	−1	−1	−1
2	−1	−1	−1	−1	−1	1	−1	34	1	−1	−1	−1	−1	1	1
3	−1	−1	−1	−1	1	−1	−1	35	1	−1	−1	−1	1	−1	1
4	−1	−1	−1	−1	1	1	1	36	1	−1	−1	−1	1	1	−1
5	−1	−1	−1	1	−1	−1	−1	37	1	−1	−1	1	−1	−1	1
6	−1	−1	−1	1	−1	1	1	38	1	−1	−1	1	−1	1	−1
7	−1	−1	−1	1	1	−1	1	39	1	−1	−1	1	1	−1	−1
8	−1	−1	−1	1	1	1	−1	40	1	−1	−1	1	1	1	1
9	−1	−1	1	−1	−1	−1	−1	41	1	−1	1	−1	−1	−1	1
10	−1	−1	1	−1	−1	1	1	42	1	−1	1	−1	−1	1	−1
11	−1	−1	1	−1	1	−1	1	43	1	−1	1	−1	1	−1	−1
12	−1	−1	1	−1	1	1	−1	44	1	−1	1	−1	1	1	1
13	−1	−1	1	1	−1	−1	1	45	1	−1	1	1	−1	−1	−1
14	−1	−1	1	1	−1	1	−1	46	1	−1	1	1	−1	1	1
15	−1	−1	1	1	1	−1	−1	47	1	−1	1	1	1	−1	1
16	−1	−1	1	1	1	1	1	48	1	−1	1	1	1	1	−1
17	−1	1	−1	−1	−1	−1	−1	49	1	1	−1	−1	−1	−1	1
18	−1	1	−1	−1	−1	1	1	50	1	1	−1	−1	−1	1	−1
19	−1	1	−1	−1	1	−1	1	51	1	1	−1	−1	1	−1	−1
20	−1	1	−1	−1	1	1	−1	52	1	1	−1	−1	1	1	1
21	−1	1	−1	1	−1	−1	1	53	1	1	−1	1	−1	−1	−1
22	−1	1	−1	1	−1	1	−1	54	1	1	−1	1	−1	1	1
23	−1	1	−1	1	1	−1	−1	55	1	1	−1	1	1	−1	1
24	−1	1	−1	1	1	1	1	56	1	1	−1	1	1	1	−1
25	−1	1	1	−1	−1	−1	1	57	1	1	1	−1	−1	−1	−1
26	−1	1	1	−1	−1	1	−1	58	1	1	1	−1	−1	1	1
27	−1	1	1	−1	1	−1	−1	59	1	1	1	−1	1	−1	1
28	−1	1	1	−1	1	1	1	60	1	1	1	−1	1	1	−1
29	−1	1	1	1	−1	−1	−1	61	1	1	1	1	−1	−1	1
30	−1	1	1	1	−1	1	1	62	1	1	1	1	−1	1	−1
31	−1	1	1	1	1	−1	1	63	1	1	1	1	1	−1	−1
32	−1	1	1	1	1	1	−1	64	1	1	1	1	1	1	1

A.1.3.2 2^{6-1} Fractional Factorial Design

No.	X_1	X_2	X_3	X_4	X_5	X_6	No.	X_1	X_2	X_3	X_4	X_5	X_6
1	−1	−1	−1	−1	−1	−1	17	1	−1	−1	−1	−1	1
2	−1	−1	−1	−1	1	1	18	1	−1	−1	−1	1	−1
3	−1	−1	−1	1	−1	1	19	1	−1	−1	1	−1	−1
4	−1	−1	−1	1	1	−1	20	1	−1	−1	1	1	1
5	−1	−1	1	−1	−1	1	21	1	−1	1	−1	−1	−1
6	−1	−1	1	−1	1	−1	22	1	−1	1	−1	1	1
7	−1	−1	1	1	−1	−1	23	1	−1	1	1	−1	1
8	−1	−1	1	1	1	1	24	1	−1	1	1	1	−1
9	−1	1	−1	−1	−1	1	25	1	1	−1	−1	−1	−1
10	−1	1	−1	−1	1	−1	26	1	1	−1	−1	1	1
11	−1	1	−1	1	−1	−1	27	1	1	−1	1	−1	1
12	−1	1	−1	1	1	1	28	1	1	−1	1	1	−1
13	−1	1	1	−1	−1	−1	29	1	1	1	−1	−1	1
14	−1	1	1	−1	1	1	30	1	1	1	−1	1	−1
15	−1	1	1	1	−1	1	31	1	1	1	1	−1	−1
16	−1	1	1	1	1	−1	32	1	1	1	1	1	1

A.1.3.3 2^{5-1} Fractional Factorial Design: Table 3.7

A.1.3.4 2^{4-3} Saturated Fractional Factorial Design

No.	X_1	X_2	X_3	X_4	X_5	X_6	X_7	X_8	X_9	X_{10}	X_{11}	X_{12}	X_{13}	X_{14}	X_{15}
1	−1	−1	−1	−1	1	−1	−1	−1	−1	1	1	1	1	1	1
2	−1	−1	−1	1	−1	−1	1	1	1	1	1	−1	1	−1	−1
3	−1	−1	1	−1	−1	1	−1	1	1	1	−1	1	−1	1	−1
4	−1	−1	1	1	1	1	1	−1	−1	1	−1	−1	−1	−1	1
5	−1	1	−1	−1	−1	1	1	−1	1	−1	1	1	−1	−1	1
6	−1	1	−1	1	1	1	−1	1	−1	−1	1	−1	−1	1	−1
7	−1	1	1	−1	1	−1	1	1	−1	−1	−1	1	1	−1	−1
8	−1	1	1	1	−1	−1	−1	−1	1	−1	−1	−1	1	1	1
9	1	−1	−1	−1	−1	1	1	1	−1	−1	−1	−1	1	1	1
10	1	−1	−1	1	1	1	−1	−1	1	−1	−1	1	1	−1	−1
11	1	−1	1	−1	1	−1	1	−1	1	−1	1	−1	−1	1	−1
12	1	−1	1	1	−1	−1	−1	1	−1	−1	1	1	−1	−1	1
13	1	1	−1	−1	1	−1	−1	1	1	1	−1	−1	−1	−1	1
14	1	1	−1	1	−1	−1	1	−1	−1	1	−1	1	−1	1	−1
15	1	1	1	−1	−1	1	−1	−1	−1	1	1	−1	1	−1	−1
16	1	1	1	1	1	1	1	1	1	1	1	1	1	1	1

A.1.3.5 2^{3-2} Saturated Fractional Factorial Design

No.	X_1	X_2	X_3	X_4	X_5	X_6	X_7
1	−1	−1	−1	−1	1	1	1
2	−1	−1	1	1	1	−1	−1
3	−1	1	−1	1	−1	1	−1
4	−1	1	1	−1	−1	−1	1
5	1	−1	−1	1	−1	−1	1
6	1	−1	1	−1	−1	1	−1
7	1	1	−1	−1	1	−1	−1
8	1	1	1	1	1	1	1

A.1.4 3^n FACTORIAL DESIGN

A.1.4.1 3^2 Factorial Design

No.	X_1	X_2
1	−1	−1
2	−1	1
3	1	−1
4	1	1
5	−1	0
6	1	0
7	0	−1
8	0	1
9	0	0

A.1.4.2 3^3 Factorial Design: Table 3.8

A.1.4.3 3^4 Factorial Design

No.	X_1	X_2	X_3	X_4	No.	X_1	X_2	X_3	X_4	No.	X_1	X_2	X_3	X_4
1	−1	−1	−1	−1	28	−1	0	1	1	55	0	1	0	−1
2	−1	−1	−1	1	29	1	0	−1	−1	56	0	1	0	1
3	−1	−1	1	−1	30	1	0	−1	1	57	0	−1	−1	0
4	−1	−1	1	1	31	1	0	1	−1	58	0	−1	1	0
5	−1	1	−1	−1	32	1	0	1	1	59	0	1	−1	0
6	−1	1	−1	1	33	−1	−1	0	−1	60	0	1	1	0
7	−1	1	1	−1	34	−1	−1	0	1	61	−1	0	0	−1
8	−1	1	1	1	35	−1	1	0	−1	62	−1	0	0	1

(continued)

(Continued)

9	1	−1	−1	−1	36	−1	1	0	1	63	1	0	0	−1
10	1	−1	−1	1	37	1	−1	0	−1	64	1	0	0	1
11	1	−1	1	−1	38	1	−1	0	1	65	−1	0	−1	0
12	1	−1	1	1	39	1	1	0	−1	66	−1	0	1	0
13	1	1	−1	−1	40	1	1	0	1	67	1	0	−1	0
14	1	1	−1	1	41	−1	−1	−1	0	68	1	0	1	0
15	1	1	1	−1	42	−1	−1	1	0	69	−1	−1	0	0
16	1	1	1	1	43	−1	1	−1	0	70	−1	1	0	0
17	0	−1	−1	−1	44	−1	1	1	0	71	1	−1	0	0
18	0	−1	−1	1	45	1	−1	−1	0	72	1	1	0	0
19	0	−1	1	−1	46	1	−1	1	0	73	0	0	0	−1
20	0	−1	1	1	47	1	1	−1	0	74	0	0	0	1
21	0	1	−1	−1	48	1	1	1	0	75	0	0	−1	0
22	0	1	−1	1	49	0	0	−1	−1	76	0	0	1	0
23	0	1	1	−1	50	0	0	−1	1	77	0	−1	0	0
24	0	1	1	1	51	0	0	1	−1	78	0	1	0	0
25	−1	0	−1	−1	52	0	0	1	1	79	−1	0	0	0
26	−1	0	−1	1	53	0	−1	0	−1	80	1	0	0	0
27	−1	0	1	−1	54	0	−1	0	1	81	0	0	0	0

A.1.4.4 Mixed $2^1 3^2$ Factorial Design

No.	X_1	X_2	X_3
1	−1	−1	−1
2	−1	1	−1
3	1	−1	−1
4	1	1	−1
5	−1	−1	1
6	−1	1	1
7	1	−1	1
8	1	1	1
9	−1	0	−1
10	1	0	−1
11	−1	0	1
12	1	0	1
13	0	−1	−1
14	0	1	−1
15	0	−1	1
16	0	1	1
17	0	0	−1
18	0	0	1

A.1.5 CENTRAL COMPOSITE DESIGN (CCD)

A.1.5.1 CCD with Two Variables: Table 3.10

A.1.5.2 CCD with Three Variables: Table 3.18

A.1.5.3 CCD with Four Variables: Table 3.27

A.2 EVOLUTIONARY OPERATION (EVOP)

A.2.1 STANDARD WORK SHEET OF EVOP FOR TWO VARIABLES A AND B

		Phase:
(diagram) Cycle: $m =$		Date:
Response:		Page:

Calculation of Averages		Standard Deviation
Operating Conditions (1) (2) (3) (4) (5)		
(i) Previous cycle sum		Previous sum $S =$
(ii) Previous cycle average		Previous average $S =$
(iii) New observations		New $S =$ Range $\cdot f_{5,m}$ $=$
(iv) Differences [(ii)−(iii)]		Range of (iv) $=$
(v) New sums [(i)+(iii)]		New sum S $=$
(vi) New averages [yi = (v)/n]		New average S $=$ new sum $S/(n-1)$

Calculation of Effects	Calculation of Error Limits
Variable A effect $= 1/2(\bar{Y}_3 + \bar{Y}_4 - \bar{Y}_2 - \bar{Y}_5)$	For new average $2S/\sqrt{n}$ $=$
Variable B effect $= 1/2(\bar{Y}_3 + \bar{Y}_5 - \bar{Y}_2 - \bar{Y}_4)$	For new effects $2S/\sqrt{n}$
$A \cdot B$ interaction effect $= 1/2(\bar{Y}_2 + \bar{Y}_3 - \bar{Y}_4 - \bar{Y}_5)$	
Change-in-mean effect $= 1/5(\bar{Y}_2 + \bar{Y}_3 + \bar{Y}_4 + \bar{Y}_5 - 4 \cdot \bar{Y}_1)$	For change in mean $1.78 S/\sqrt{n} =$

A.2.2 VALUES OF $f_{k,m}$ IN EVOP

k \ m	2	3	4	5	6	7	8	9	10
5	0.30	0.35	0.37	0.38	0.39	0.40	0.40	0.40	0.41
9	0.24	0.27	0.29	0.30	0.31	0.31	0.31	0.32	0.32
10	0.23	0.26	0.28	0.29	0.30	0.30	0.30	0.31	0.31

A.3 LIST OF COMPUTER PROGRAMMS

- SCREEN.XLS: MS-Excel® program to analyze screening data (Figure 3.1)
- ALPHA.XLS: MS-Excel® program to calculate the α value (Figure 3.9)
- REG_SPSS.INC: SPSS/PC+ macro for regression analysis (Table 3.13)
- REG_SAS.SAS: SAS macro for regression analysis (Table 3.14)
- S_TIU01.INC: SPSS/PC+ macro for modeling of inactivation degree of trypsin inhibitor
- S_TIU01.DAT: Experimental data of inactivation degree of trypsin inhibitor (Table 3.18)
- S_TIU01.LIS: Modeling results of inactivation degree of trypsin inhibitor (Table 3.19)
- X-Y-PLOT.XLS: MS-Excel® program to generate 2D X-Y-plots (Section 3.7.1.1)
- SURFACE.XLS: MS-Excel® program to generate response surface plots (Section 3.7.1.2)
- 2-3D_SYS.CMD: SYSTAT macro to generate 2D+3D response surface plots (Table 3.20)
- 3SURFACE.DRW: A 3D contour surface plot (Figure 3.28)
- EVOP.DOC: Standard worksheet of EVOP
- OPTIMUM1.BAS: BASIC program doing optimization (Table 3.30)
- OPTIMUM1.EXE: Executable optimization program, compiled from OPTIMUM1.BAS
- OPTIMUM1.DAT: Optimization results (Table 3.31)
- REG_SASM.SAS: SAS macro for regression analysis of mixture data (Table 4.9)
- REG_SASM.LOG: Modeling results using SAS (Table 4.10)
- TRACEPLT.XLS: MS-Excel® program to generate trace plots (Section 4.7.1)
- M_CONOUR.BAS: BASIC program to generate 2D simplex contour plots (Table 4.11)
- M_CONOUR.EXE: Executable program to generate 2D simplex contour plots
- M_OPTM1.BAS: BASIC program doing optimization of a mixture system with two objective responses (Table 4.13)
- OPTIM.DAT: Optimization results of running M_OPTIM1.BAS (Table 4.14)
- M_MODEL.XLS: MS-Excel® program to calculate the coefficients in a combined model (Figure 5.5)

A.4 SCHEMA OF SENSORY EVALUATION OF INSTANT FOOD

1. Appearance, Surface, Particle Size, Color						Evaluation 5 4 3 2 1 0 × 6 =						
• unequal particle size	4	3	2		-	**Color**						
• impure	4	3			-	• weak	4	3	2		-	
• husk	4	3	2		-	• intensive	4	3	2		-	
• piece of hull	4	3	2		-	• irregular	4	3	2		-	
						• faulty color	4	3	2		-	
Surface						**Other faults**						
• not homogeneous	4	3			-	• not evaluable	4	3	2	1	-	
• stained, blotchy	4	3			-		-	-	-	-	0	

2. Texture, Viscosity						Evaluation 5 4 3 2 1 0 × 5 =						
• thick	4	3			-	**Other faults***						
• thin	4	3			-	• not evaluable	4	3	2	1	-	
• clump	4	3			-		-	-	-	-	0	

3. Odor, Taste						Evaluation 5 4 3 2 1 0 × 6 =						
• lack of aroma	4	3	2		-	• impure odor	4	3	-	-	-	
• bitter	4	3			-	• strange odor	-	-	2	1	-	
• dull	4	3			-	• impure taste	4	3	-	-	-	
• scratched, old	4	3			-	• strange taste	-	-	2	1	-	
• greasy, oily	4	3	2		-							
• rancid	4	3			-	Other faults						
• musty, muffled	4	3			-	• not evaluable	4	3	2		-	
	4	3	2		-		-	-	-	-	0	

Preparation of samples	Notices about other faults:	Weighted Total Score / Sum of Weight Factor × 1/20 = [ ]
☐ Dry ☐ Watery ☐ Moist ☐ Slurry		Quality Score

Index